MINITAB Laboratory Manual

to accompany

Regression Analysis: Concepts and Applications

MINITAB Laboratory Manual

to accompany

Regression Analysis: Concepts and Applications

Franklin A. Graybill
Colorado State University

Hariharan K. Iyer
Colorado State University

Duxbury Press
An Imprint of Wadsworth Publishing Company
Belmont, California

Duxbury Press
An Imprint of Wadsworth Publishing Company
A Division of Wadsworth, Inc.

I(T)P™
International Thomson Publishing
The trademark ITP is used under license.

Printed in the United States of America

1 2 3 4 5 6 7 8 9 10—98 97 96 95 94

Contents

Preface

This laboratory manual is designed to be used as a companion to *Regression Analysis: Concepts and Applications* which we refer to as the textbook. Anyone acquainted with regression analysis knows that it is difficult, and practically speaking impossible, to do the computing required in multiple regression analysis without the use of a statistical computing package. In this companion manual we explain how to use the statistical package Minitab to do the computations needed in the textbook.

We chose Minitab because it is very easy to learn and yet sufficiently general and supports a variety of statistical procedures. Many universities use Minitab as a tool for teaching statistics. It is particularly suitable for a course in multiple regression analysis.

Many simple Minitab commands are available for doing the computing required for most of the regression techniques discussed in the textbook. However, there are a few regression procedures for which no simple commands exist in the present version of Minitab to do the required computations. For these, we have written Minitab programs, referred to as macros, that can be used, and these macros are on the disk that accompanies this manual. The macros are meant to be used for problems and exercises in the textbook and are not meant for general purpose applications.

Although none of the data sets used in the textbook are "real" data sets, they are "real-appearing" and they do not depend upon detailed knowledge of any particular field such as biology, business, chemistry, economics, engineering, physics, etc. The data sets involve common, day-to-day topics such as **insurance premiums, grade point averages, final and midterm exams, electric bills, grocery costs, strength of plastic containers, professors' salaries, drug concentration in blood, effect of haze on visibility, age and blood pressure**, etc.

All of the data used in the textbook are on the disk that accompanies this manual. Each data file is stored in two formats – as an ASCII file and also as a portable Minitab worksheet file. The ASCII files have the extension **dat** whereas the Minitab worksheet files have the extension **mtp**. Except for the data files which involve entire population data, none of the files have observation numbers included as part of the dataset. The first column is always reserved for the response variable and the subsequent columns are reserved for predictor variables (there are one or two exceptions to this general rule and these will be pointed out where appropriate). The Minitab macros are written assuming this arrangement of the response variable and the predictor variables. The Minitab worksheet files have the data as well as variable names and other useful information but the ASCII files have only the data.

This manual should not be considered as a tutorial for Minitab, but as a companion manual to the textbook and as an aid to studying regression. Often there may be alternative Minitab commands that can be used to perform a particular series of computations, and the ones we use may not be the best or the most efficient. However, we chose them as a simple way to do the computations without detracting from the main subject which is regression. Those of you who are interested in learning more about Minitab should consult the *Minitab Reference Manual* and the *Minitab Handbook.*

Chapter 1

Review of Basic Statistical Concepts and Matrices

1.1 Overview

This laboratory manual explains how to use the statistical package Minitab to perform calculations for each procedure discussed in the book *Regression Analysis: Concepts and Applications*, which we refer to as the textbook.

General Comments

The sections in this manual correspond to sections in the textbook. For example, Section 4.6 in this manual corresponds to Section 4.6 in the textbook, etc. If any problem or exercise in the textbook requires computations, this laboratory manual explains how to do those computations using the statistical package Minitab. Whenever we refer to a chapter, a section, an equation number, a table, a figure, an exhibit, or a box, these will be the chapter, section, equation number, etc., in the textbook unless we explicitly refer to this manual. Tables in the laboratory manual are named Table A, Table B, etc. For example, Table A refers to a table in this manual.

What Is Minitab?

Minitab is a general purpose statistical computing pacakge. It is easy to use and yet powerful enough for many data analysis situations. It is widely available for PCs as well as mainframe computers.

Early versions of Minitab were based on a command language, but newer versions allow analysts the option of using the command language or using a mouse and a menu system. Typically, a user reads a set of data into Minitab, then types appropriate commands. When a user types a command from the keyboard and presses the `Enter` key, Minitab responds by displaying the results of the calculations it was asked to perform.

For our discussion we assume that you are working on a personal computer that has a hard disk drive, usually called the `C` drive, and at least one floppy disk drive, say the `B` drive. The data disk that accompanies this manual should be inserted in drive `B`. We assume that the subdirectory where Minitab is installed is declared in an appropriate *path* statement in your *autoexec.bat* file. If you have no prior experience with personal computers, you should seek help from your laboratory instructor.

Invoking and Exiting Minitab

We assume your current working directory is `C:\`. If so, the computer will display the prompt `C:\>`. On most systems you can start Minitab by typing the word `minitab` and pressing the `Enter` key. Minitab will then display the version number you are using and some other introductory comments and finally display the prompt `MTB >`. This prompt tells you that Minitab is expecting you to type in a command. To exit Minitab, type `stop` and press `Enter`.

How Minitab Stores Data

Data are read into Minitab from data files (such as those supplied on the data disk accompanying this manual) or by entering them using the keyboard. Minitab stores these data in a worksheet. You can think of this worksheet as consisting of rows and columns. The columns have the default names `C1`, `C2`, ..., etc. The label `C1` refers to column 1 of the worksheet, the label `C2` to column 2 of the worksheet, and so on. You can also specify these columns using lower case letters, e.g., `c1`, `c2`, ..., etc., since Minitab does not distinguish between lower and upper case for column names. The total number of rows and columns in the worksheet available for data storage depends on the particular PC or mainframe system you are using, but typically adequate space is available for reading in and storing any data set you may work with from the textbook.

The values for any variable of interest are typically stored as elements of one of the columns of the worksheet. For example, if you are interested in working with heights and weights of a group of people, then you can store the heights in column `c1` and the weights in column `c2`. Each row of data (in columns `c1` and `c2`) will correspond to one of the people in this group. You can access the data items from any row and any column using appropriate commands. These will be discussed later.

Minitab also has special memory locations in the worksheet for storing single numbers (referred to as *constants* by Minitab). These storage locations are labeled `K1`, `K2`, ..., etc. Additionally, there are storage locations in the worksheet for storing *matrices* that are labeled `M1`, `M2`, ..., etc. The number of *constants* and *matrices* that can be stored depends, again, on the amount of memory available in your particular system. You will learn more about constants and matrices in the next few sections.

Many useful statistical calculations can be carried out with relatively few commands using Minitab. For example, suppose you have data in a column, say `c1` (column 1). Then the simple command

```
describe c1
```

will compute the following:

- The mean of the data
- The standard deviation of the data
- The minimum and maximum values of the data
- The median of the data
- Many other useful quantities

Thus you can see that one simple command will compute several quantities. To use Minitab for analyzing a set of data you must

(a) Start the Minitab program

(b) Enter the data into the computer

(c) Perform the required calculations or obtain the required plots

(d) Output the results so you can read them or save them in a file for later use.

We presume that **(a)** has been accomplished. Here we discuss **(b)**.

Reading Data into the Minitab Worksheet

Data may be entered into the Minitab worksheet in one of the following ways:

(1) By entering the data via the computer keyboard

(2) By reading the data from a data file (for instance, the files on the data disk accompanying this manual or your own data files)

First we show you how to enter the data via the keyboard.

Entering Data via the Keyboard

After you get into the Minitab system, you will see the Minitab prompt

```
MTB >
```

On the same line as the prompt, type the necessary instructions and then press the `Enter` key. These instructions are referred to as Minitab commands. Suppose you want to perform some computations using the data in Table A below.

Table A

Y	X
1.2	6.0
1.8	6.3
2.8	5.8
2.7	5.7
3.5	6.4

You must first get the data into the Minitab worksheet. The instructions below provide one way of entering the data in Table A into the Minitab worksheet via the keyboard.

(1) Following the Minitab prompt

```
MTB >
```

you type `set c1` and press the `Enter` key. This tells the computer that you are going to enter data and put it in column `c1` (instead of column `c1` you can use any other column, for example column `c23`). As stated earlier, the number of columns you can use depends on the version of Minitab you are using and the memory available in your computer.

(2) After you type `set c1` and press `Enter`, the prompt

```
DATA>
```

appears, which is your cue to enter the data that will go into column `c1`. You can enter data by typing in one number at a time and pressing `Enter` after each number is typed in, or, you can enter data by typing as many data values as will fit on each row and then pressing `Enter` at the end of the row, continuing to type in data values on the next row if necessary. If several data values are entered in a row, then there must be at least one space between each datum that you enter. For example, you can enter the five numbers 1.2, 1.8, 2.8, 2.7, and 3.5 in Table A as

```
1.2
1.8
2.8
2.7
3.5
```

or as

```
1.2 1.8 2.8 2.7 3.5
```

but not as

```
1.21.82.82.73.5
```

You can use as many lines as necessary to type in the data values. After each line of data values, press `Enter` and then continue to type in data values on the next line. After typing all the data that you want to put into column `c1`, press the `Enter` key and type `end`. Until you type `end`, Minitab will keep giving you the prompt

```
DATA>
```

reminding you that more data values are expected. After you type `end`, signifying the end of data entry, you will get the prompt

```
MTB >
```

at which point you can type any other Minitab command.

The commands and prompts for entering the Y data in Table A via the keyboard are shown below. The capital letters preceding the > symbol (including this symbol) are part of the Minitab prompt. The Minitab response to the commands is given between dashed lines.

COMMAND FOR ENTERING DATA VIA THE KEYBOARD

```
MTB > set c1
DATA> 1.2 1.8 2.8 2.7 3.5
DATA> end
```

You have now entered the Y data into column c1, and you are ready to enter the X data into another column, say c2. The Minitab commands for this are given below.

COMMAND FOR ENTERING DATA VIA THE KEYBOARD

```
MTB > set c2
DATA> 6.0 6.3 5.8 5.7 6.4
DATA> end
```

You have now entered into the Minitab worksheet all the data that are in Table A above. You should print them on the screen to examine them and check to be sure that you didn't make a mistake. The Minitab command to print the data is

PRINT COMMAND

```
MTB > print c1 c2
```

Minitab responds with:

```
----------------------------------------------------------------

ROW     C1     C2

  1    1.2    6.0
  2    1.8    6.3
  3    2.8    5.8
  4    2.7    5.7
  5    3.5    6.4
----------------------------------------------------------------
```

The two columns of data are printed along with an additional column (the first one) labeled `ROW` to help you quickly locate data in any specified row. For example, the data value in `ROW 3` of column `c1` is 2.8. If you only want column `c1` to be printed, then the command is

PRINT COMMAND

```
MTB > print c1
```

Minitab responds with:

```
----------------------------------------------------------------

C1
   1.2     1.8     2.8     2.7     3.5
----------------------------------------------------------------
```

Note that when Minitab prints a single column of data, it prints the data values as a row rather than as a column in order to save space. Note that no `ROW` numbers are printed either.

Reading Data into the Minitab Worksheet from a Data File

We now show you how to read data into the Minitab worksheet from a data file that is stored on a disk. This is particularly useful since most of the data sets you will be working with are stored on the data disk accompanying this manual.

For convenience, each set of data used in the textbook has been stored in two formats on the disk that accompanies this manual. The first format is an ASCII (**A**merican **S**tandard **C**ode for **I**nformation **I**nterchange) file that contains data, and these data can easily be read into the computer by using almost any software package. The file names for ASCII files have the extension **dat**. The second format is a Minitab worksheet file that contains data and additional information such as the names of the variables and the number of observations, etc. The names for Minitab worksheet files that are on the disk that accompanies this manual have the extension **mtp**. Thus the file name **table161.dat** refers to an ASCII file that contains a data set, whereas the file name **table161.mtp** refers to a file containing the same data set along with the names of variables, number of observations, etc., stored as a Minitab worksheet file.

As standard practice, when you plan to analyze a new data set using Minitab, you should read the data into the Minitab worksheet, name the various columns of data, and then save them in a Minitab worksheet file. The data should always be printed and a preliminary examination of the data carried out. This examination may consist of looking at the contents of the data file, the calculation of various means and standard deviations, and plotting histograms, scatterplots, etc. This will help you become familiar with the data set with which you will be working.

ASCII Data Files

We begin by showing you how to read an ASCII data file into the Minitab worksheet, name the variables in the file, and create a Minitab worksheet file for later use. The disk that accompanies this manual contains an ASCII data file named **table161.dat**, and we use this file to show you how to create the Minitab worksheet file **table161.mtp** (these data are in Table 1.6.1). This same procedure can be used to read in any other ASCII data file (perhaps your own data files) and create and save these data in a Minitab worksheet file. The commands that follow the Minitab prompt `MTB >` are the input commands that Minitab will execute. After typing in a command, you must press the `Enter` key on the keyboard. As usual, the Minitab responses are enclosed between dashed lines.

The ASCII data file **table161.dat** contains one column of data (the car numbers that are shown in Table 1.6.1 are not contained in the data file). We read this file and put the data in column `c1` using the command below. Note that both quotes enclosing the file name in the command are right quotes ’.

READ COMMAND

```
MTB > read 'b:\table161.dat' c1
```

Minitab responds with:

```
------------------------------------------------------------------------

     10 ROWS READ

C1
  25.72   25.24   25.19   25.88   .   .   .
------------------------------------------------------------------------
```

Note that we have used lower case letters for the file name. DOS systems are case insensitive, so whether you use upper case letters or lower case letters or a mixture of upper case and lower case letters is immaterial. But UNIX systems and some other systems are case sensitive and you must pay particular attention to upper and lower case.

We recommend that you assign a name to each column of data for convenience of reference. To illustrate, we give the name `mpg` to the data in column `c1` since it contains the miles per gallon for each of the cars. To name the column of data, we give the following Minitab command.

COMMAND TO NAME A COLUMN

```
MTB > name c1 'mpg'
```

The `read` command can also be used to enter data into the Minitab worksheet from the keyboard, especially for entering multiple columns of data. For example, to enter the two columns of data given in Table A you can use the following command.

ENTERING DATA VIA THE KEYBOARD USING THE READ COMMAND

```
MTB > read c1 c2
DATA> 1.2 6.0
DATA> 1.8 6.3
DATA> 2.8 5.8
DATA> 2.7 5.7
DATA> 3.5 6.4
DATA> end
```

Minitab responds with:

```
--------------------------------------------------------------------------

      5 ROWS READ
--------------------------------------------------------------------------
```

Alternatively, you can enter each column of data seperately by using the `set` command as discussed earlier.

The ERASE Command

The `erase` command may be used to erase the contents of specified columns of the worksheet. For instance, suppose that columns `c1` and `c2` contain the Y and the X values, respectively, from Table A. To erase the contents of column `c2` of the worksheet, the command is

```
erase c2
```

After this command is executed, only column `c1` will have data values, viz., the Y values from Table A.

Saving the Contents of the Minitab Worksheet

The `save` command saves the data in the Minitab worksheet in the form of a Minitab worksheet file along with the names of the columns, and it allows the data to be retrieved during a later Minitab session by using the `retrieve` command. To save the contents of a worksheet in a Minitab worksheet file that can be retrieved and used later (perhaps even on a different computer system), you must use a multiline command. The first line of a multiline command is called the **main command**, and all remaining lines are called **subcommands**. The format of a general multiline command is shown below.

MULTILINE MINITAB COMMAND

```
MTB > main command;
SUBC> first subcommand;
SUBC> second subcommand;
SUBC> ....;
SUBC> ....;
SUBC> last subcommand.
```

Thus in a multiline Minitab command, each line except the last line ends with a semi-colon, whereas the last line must end with a period.

To save the contents of the Minitab worksheet in a Minitab worksheet file named **table161.mtp** in the subdirectory `c:\mtbwork` (assuming that this subdirectory already exists), the command and subcommand are

SAVE COMMAND AND PORTABLE SUBCOMMAND

```
MTB > save 'c:\mtbwork\table161';
SUBC> portable.
```

Minitab responds with:

```
------------------------------------------------------------------

Worksheet saved into file: c:\mtbwork\table161.MTP
------------------------------------------------------------------
```

Thus Minitab's response shows that the contents of the Minitab worksheet have been saved in a Minitab worksheet file named `table161.MTP` in the subdirectory `c:\mtbwork`. Note that Minitab automatically adds the file extension `MTP` (Minitab uses upper case letters for `MTP` but we are free to use lower case or upper case when referring to it) to the file name `table161` and we do not need to explicitly include the extension `mtp` or `MTP` in the `save` command. The contents of the worksheet are saved in a form that is very similar to an ASCII data file, but in addition to the data, the names you have given to columns are also saved; also, all constants and matrices that are part of the Minitab worksheet are saved. We say more about this later. If you want to save the file **table161.mtp** in a subdirectory other than `c:\mtbwork`, then you must use the appropriate path name in place of `c:\mtbwork`.

You can give a file any name as long as it obeys the rules for naming a file. In this manual we use file names which have two parts separated by a period . where the name before the period consists of a combination of no more than eight letters and numbers. These eight letters and numbers are typically chosen to help you remember the contents of the file. For example, the filename **tableA.mtp** reminds us that the data stored in this file are the data given in Table A. The name after the period, which is often referred to as the **file extension**, can contain no more than three letters, and identifies the type of file. For example, the extension **mtp** to a file name tells us that it is a **portable** (the **p** in **mtp** stands for **portable**) Minitab worksheet file. A portable file is a file that can be read into the Minitab worksheet by many different computer systems.

The file **table161.mtp** can be retrieved at any time during a Minitab session by giving the folowing command and subcommand (we are assuming that the subdirectory `c:\mtbwork` is where you stored the file **table161.mtp**).

COMMAND AND SUBCOMMAND TO RETRIEVE A PORTABLE FILE

```
MTB > retrieve 'c:\mtbwork\table161';
SUBC> portable.
```

Minitab responds with:

```
------------------------------------------------------------------

Worksheet retrieved from file: c:\mtbwork\table161.MTP

------------------------------------------------------------------
```

Thus Minitab tells us that the file **table161.MTP** has now been retrieved and is in the main memory of the computer. Note that we do not need to specify the file extension MTP explicitly.

The INFO Command

The command `info` (short for `information`) gives a summary of the contents of various columns of data and any constants or matrices that are part of the worksheet. To illustrate, first retrieve the file **table161.mtp** from the data disk using the following command and subcommand.

```
retrieve 'b:\table161';
portable.
```

To obtain information about the contents of this file, issue the following command.

INFORMATION COMMAND

```
MTB > info
```

Minitab responds with:

```
------------------------------------------------------------------

COLUMN    NAME      COUNT
C1        mpg          10

CONSTANTS USED: NONE
------------------------------------------------------------------
```

We observe that the worksheet currently contains one column of data, c1, and that this column consists of 10 (under COUNT) numbers. The name of the variable associated with the column of numbers is given under the column labeled NAME. Note the words

```
CONSTANTS USED: NONE
```

which tell us that no constants are saved in the worksheet.

The DESCRIBE Command

The **describe** command is used to obtain summary statistics for specified columns of data. For an illustration, first retrieve the file **table161.mtp** from the data disk. To obtain summary statistics for column c1 the command is

DESCRIBE COMMAND

```
MTB > describe c1
```

Minitab responds with:

```
--------------------------------------------------------------------------

             N      MEAN    MEDIAN    TRMEAN     STDEV    SEMEAN
mpg         10    25.302    25.215    25.289     0.639     0.202

           MIN       MAX        Q1        Q3
mpg     24.290    26.420    24.915    25.760
--------------------------------------------------------------------------
```

The summary includes the mean and the standard deviation. It also includes the median, the maximum and the minimum of the numbers, and other useful quantities. You should consult the *Minitab Reference Manual* for details.

> **Note:** You should be aware that Minitab always calculates standard deviations using the formula
>
> $$\sqrt{\frac{1}{n-1}\sum_{i=1}^{n}(Y_i - \overline{Y})^2}$$
>
> which corresponds to (1.6.2).

The WRITE Command

Sometimes it is necessary to save the data from selected columns of the Minitab worksheet as an ASCII file so that these data may be used with other computer packages. This can be done by the use of the `write` command. For illustration, suppose that the Minitab worksheet contains several columns of data, and that you want to save the data contained in columns c1 and c2 in a file named **mydata.dat** and store it in the subdirectory `c:\mtbwork` . The required Minitab command for this is

WRITE COMMAND

```
MTB > write 'c:\mtbwork\mydata.dat' c1 c2
```

With this command we have placed the data that are in columns `c1` and `c2` into a file to which we have given the name `mydata.dat`. This file is saved in the subdirectory `c:\mtbwork` (if no path is specified for the file **mydata.dat** then, by default, Minitab will attempt to use the current working directory). Note that the column names are not saved as part of this file since it is a **dat** file. Only the data values are saved.

For additional commands available in Minitab you can consult the *Minitab Reference Manual.* Recent versions of Minitab have online help available to you. Type the command `help` and press `Enter` to learn more about available commands in Minitab.

Problems

For each problem, give the appropriate Minitab commands and the answer when required. Problems **M1.1.1–M1.1.6** refer to the data below.

Y	X	Z
1.5	600	34.5
1.9	590	43.9
1.2	710	30.3
2.1	560	31.7
1.6	610	42.1
1.7	700	39.4

M1.1.1 Enter the data into the Minitab worksheet via the keyboard. Put Y values in column `c1`, X values in column `c2`, and Z values in column `c3`.

M1.1.2 Give column `c1` the name Y, column `c2` the name X, and column `c3` the name Z.

M1.1.3 Print the data and make sure you have entered them correctly.

M1.1.4 Write these data to an ASCII data file named **prob111.dat** in your working directory on the `C` drive.

M1.1.5 Save these data in a Minitab worksheet file named **prob111.mtp** in your working directory on the `C` drive.

M1.1.6 Use the `describe` command to find the mean of Y, the standard deviation of X, and the maximum and minimum of Z. Check these by hand calculations.

Problems **M1.1.7–M1.1.11** refer to the data in the file **table161.mtp** on the data disk. These data are also given in Table 1.6.1.

M1.1.7 Retrieve the data contained in the Minitab worksheet file **table161.mtp** on the data disk and use the command `info` to see what is in the file.

M1.1.8 Print the data and check the numbers.

M1.1.9. Give the name MPgal to column `c1`.

M1.1.10 Compute the mean and the standard deviation of MPgal.

M1.1.11 What are the minimum and maximum values of MPgal?

Problems **M1.1.12–M1.1.15** refer to the data in Table 1.6.4.

M1.1.12 The data for Problem 1.6.1 are in the files **table164.dat** and **table164.mtp** on the data disk. Retrieve the data.

M1.1.13 Use the `info` command to see what the file contains.

M1.1.14 Print the data and compare them with the data in Table 1.6.4 to be sure they are the same.

M1.1.15 Use the `describe` command to find the mean, the standard deviation, and the median of the data. Check the calculations by hand.

1.2 Basic Ingredients for Statistical Inference

There are no calculations in this section that require Minitab.

1.3 Populations

There are no calculations in this section that require Minitab.

1.4 Model

There are no calculations in this section that require Minitab.

1.5 Parameters (Summary Numbers)

There are no calculations in this section that require Minitab.

1.6 Samples and Inferences

In this section we explain several Minitab commands that are useful for computing various quantities discussed in Section 1.6 of the textbook.

Some Arithmetic Using Minitab

Here we show you how to perform some simple arithmetic calculations using Minitab. We use the data in Table B below for illustrating the commands.

Table B

C1	C2	C3
12	2	32
21	4	16
31	1	35
52	5	25
37	3	27
35	6	24

You can add, subtract, multiply, and divide any two columns of data (element by element) by using the Minitab command `let` . The columns must all have the same number of elements. For example, the command `let c4=c1+c2` adds the corresponding elements in columns `c1` and `c2` and puts the results in column `c4`. We illustrate below using the three columns of data in Table B. Note that the symbol `*` is used for multiplication. We first enter the data from Table B into the Minitab worksheet using the following commands.

```
set c1
12 21 31 52 37 35
set c2
2 4 1 5 3 6
```

```
set c3
32 16 35 25 27 24
```

The following commands illustrate the various arithmetic operations.

THE LET COMMANDS

```
MTB > let c4=c1+c2
MTB > let c5=c3-c2
MTB > let c6=c1/c2
MTB > let c7=c1*c2
```

Next we print c1, c2, c3, c4, c5, c6, and c7 with the command

```
MTB > print c1 c2 c3 c4 c5 c6 c7
```

Minitab responds with:

ROW	C1	C2	C3	C4	C5	C6	C7
1	12	2	32	14	30	6.0000	24
2	21	4	16	25	12	5.2500	84
3	31	1	35	32	34	31.0000	31
4	52	5	25	57	20	10.4000	260
5	37	3	27	40	24	12.3333	111
6	35	6	24	41	18	5.8333	210

You should check these results to be sure you understand what the commands do.

Sometimes we do not want to work with columns of data but rather with single numbers (often called *scalars*). Minitab denotes single numbers by K1, K2, ...(you can use lower case letters, viz., k1, k2, ..., if you wish). For example, study the following commands:

LET COMMAND

```
MTB > let k1=15
```

LET COMMAND

```
MTB > let k2=5
```

We can add, subtract, multiply, and divide k1 and k2 and put the result into another k, say k4. For example, if we want to multiply k1 by k2 and put the result in k3, the command is

LET COMMAND

```
MTB > let k3=k1*k2
```

Use the command print k1 k2 k3 to view the values of k1, k2, and k3. We can also multiply a column of data by a single number. To multiply every number in column c2 by the value of k1 and put the result in c12 the command is

LET COMMAND

```
MTB > let c12=k1*c2
```

Instead of multiplying by the value of k1 or k2, etc., we could just use the number directly. The command could be

LET COMMAND

```
MTB > let c12=15*c2
```

Note that, if in any division the divisor is zero, Minitab will tell you that you have committed an error.

Sums, Means, Sums of Squares, and Sums of Crossproducts Using Minitab

We use the data in Table B above to show you the Minitab commands for computing sums, means, sums of squares, sums of squares of deviations, and sums of crossproducts. First enter the data values from the three columns of Table B into the Minitab worksheet and put them in, say, columns c1, c2, and c3, respectively.

Suppose you want to compute the sum of the numbers in column c1 and save the result in k1. The command is

COMMAND FOR COMPUTING THE SUM OF THE NUMBERS IN A COLUMN

```
MTB > let k1=sum(c1)
```

To obtain the mean of the numbers in a column, say in column c2, and save the result in k2, the command is

COMMAND FOR COMPUTING THE MEAN OF THE NUMBERS IN A COLUMN

```
MTB > let k2=mean(c2)
```

To calculate the sum of squares, $\sum_{i=1}^{n} y_i^2$ of the numbers in column c2 and store the result in k3, the command is

COMMAND FOR COMPUTING THE SUM OF SQUARES OF THE NUMBERS IN A COLUMN

```
MTB > let k3=ssq(c2)
```

You can also use the command `let k3=sum(c2*c2)` to get the result. To compute $SSY = \sum_{i=1}^{n}(y_i - \overline{y})^2$ for the data in column c1 and store the result in k11, the command is

COMMAND FOR COMPUTING A SUM OF SQUARES OF DEVIATIONS

```
MTB > let k11=ssq(c1-mean(c1))
```

To calculate $SXY = \sum_{i=1}^{n}(y_i - \overline{y})(x_i - \overline{x})$, where y_i and x_i, $i = 1, \ldots, n$ are in columns c1 and c2, respectively, and save the result in k12, the command is

COMMAND FOR COMPUTING A SUM OF CROSSPRODUCTS

```
MTB > let k12=sum((c1-mean(c1))*(c2-mean(c2)))
```

We view the results of the above calculations by using the command

```
print k1 k2 k3 k11 k12
```

Minitab responds with:

```
---------------------------------------------------------------------

K1          188.000
K2          3.50000
K3          91.0000
K11         953.333
K12         62.0000
---------------------------------------------------------------------
```

You should check these results to be sure that you understand the commands.

Sometimes there is a need for performing arithmetic operations on selected elements of a column or columns rather than on entire columns of numbers. For instance, in Table B above, suppose we want to multiply the second element of column c1 by the third element of c2 and add the result to the fifth element of c3 and store the final result in k5. This can be done using the following command.

LET COMMAND FOR ARITHMETIC OPERATIONS ON SELECTED ELEMENTS

```
MTB > let k5=c1(2)*c2(3)+c3(5)
```

Using the command `print k5` we observe that the value of k5 is 48.0000. You should experiment with variations of this command using other operations.

Computing Confidence Intervals Using Minitab

There is a command in Minitab for computing a confidence interval for the mean μ_Y

of a population {Y} using sample data. The formula used is in (1.6.8). The command is

COMMAND FOR COMPUTING A CONFIDENCE INTERVAL FOR A MEAN

```
MTB > tinterval K c1
```

In this command, K is the confidence coefficient and the data are in column c1. Of course the data can be in any other column, in which case the appropriate column must be used in the command.

To illustrate this command we will use the data contained in the Minitab worksheet file **table161.mtp** on the data disk. We assume these data are a simple random sample from a Gaussian population with mean μ_Y and standard deviation σ_Y. First we retrieve these data and print them using the command print c1 .

Minitab responds with:

```
-------------------------------------------------------------------------

mpg
  25.72   25.24   25.19   25.88   26.42   24.48   25.11   24.29   25.06
  25.63
-------------------------------------------------------------------------
```

Now we use these data and compute an 80% confidence interval for the population mean with the command

COMMAND FOR COMPUTING AN 80% CONFIDENCE INTERVAL FOR A POPULATION MEAN

```
MTB > tinterval .80 c1
```

Minitab responds with:

```
-------------------------------------------------------------------------

              N      MEAN     STDEV  SE MEAN    80.0 PERCENT C.I.
mpg          10    25.302     0.639    0.202  (  25.022,   25.582)
-------------------------------------------------------------------------
```

From this we get $\hat{\mu}_Y = 25.302$, $\hat{\sigma}_Y = 0.639$, $SE(\hat{\mu}_Y) = 0.202$, and the confidence statement is

$$C[25.022 \le \mu_Y \le 25.582] = 0.80$$

which, of course, is the same as in Example 1.6.1.

Tests of Hypotheses

We now demonstrate Minitab commands for testing one-sided and two-sided hypotheses about the mean of a Gaussian population. For testing

$$\text{NH: } \mu_Y = q \text{ against AH: } \mu_Y \neq q$$

the command is

COMMAND FOR COMPUTING A TWO-SIDED TEST OF A MEAN

```
MTB > ttest q c1
```

This is a two-sided t-test of NH: $\mu_Y = q$ against AH: $\mu_Y \neq q$ where the sample data are in column c1. Minitab prints the P-value along with other useful statistics.

If you want a one-sided test instead of a two-sided test, you can do it but this requires a subcommand. Below is the command (and the subcommand) for obtaining the results of a one-sided test of

$$\text{NH}: \ \mu_Y \le q \text{ against AH}: \ \mu_Y > q$$

COMMAND FOR COMPUTING A ONE-SIDED T-TEST OF A MEAN

```
MTB > ttest q c1;
SUBC> alternative = +1.
```

Below is the command (and the subcommand) for obtaining the results of a one-sided test of

$$\text{NH}: \ \mu_Y \ge q \text{ against AH}: \ \mu_Y < q$$

COMMAND FOR COMPUTING A ONE-SIDED T-TEST OF A MEAN

```
MTB > ttest q c1;
SUBC> alternative = -1.
```

Thus the subcommand is `alternative = +1` if the alternative hypothesis is $\mu_Y > q$ (i.e., $\mu_Y - q > 0$), and it is `alternative = -1` if the alternative hypothesis is $\mu_Y < q$ (i.e., $\mu_Y - q < 0$).

To illustrate the above commands, consider Example 1.6.4 where we want to test

$$\text{NH}: \ \mu_Y \leq 25 \text{ against AH}: \ \mu_Y > 25$$

The data are in column `c1` of the Minitab file **table161.mtp** on the data disk, and we assume that this file has been retrieved and its contents carefully examined. The command and the subcommand for the required one-sided test are

COMMAND FOR COMPUTING A ONE-SIDED TEST OF A MEAN

```
MTB > ttest 25 c1;
SUBC> alternative = +1.
```

Note that a `+1` is used because the alternative hypothesis is $\mu_Y > 25$.

Minitab responds with:

```
-----------------------------------------------------------------------

TEST OF MU = 25.000 VS MU G.T. 25.000

               N      MEAN     STDEV   SE MEAN         T    P VALUE
mpg           10    25.302     0.639     0.202      1.49      0.085

-----------------------------------------------------------------------
```

The computed value of the test statistic (given under the label `T`) is $t_C = \text{T} = 1.49$ and the P-value is 0.085, the same numbers (except for rounding error) we obtained in Example 1.6.4.

Problems

Problems **M1.6.1–M1.6.10** refer to the data set in Table 1.6.4, which is a simple random sample of size 30 from a Gaussian population with mean μ_Y and standard deviation σ_Y. This data set is also in the file **table164.mtp** on the data disk. For each problem, give the appropriate Minitab commands in addition to the answer.

M1.6.1 Retrieve the data, give the column containing the data the name Y, and print it.

M1.6.2 Find the estimate of μ_Y.

M1.6.3 Find a 90% upper confidence bound for μ_Y.

M1.6.4 Find t_C, the computed t-value for testing

$$\text{NH}: \ \mu_Y = 4.5 \text{ against AH}: \ \mu_Y \neq 4.5$$

M1.6.5 In Problem **M1.6.4** find the P-value for the test.

M1.6.6 Find the t_C value and the P-value for testing

$$\text{NH}: \ \mu_Y \leq 5.0 \text{ against AH}: \ \mu_Y > 5.0$$

M1.6.7 Find $\sum_{i=1}^{30} y_i$ = sum of the sample Y values.

M1.6.8 Find $\sum_{i=1}^{30} y_i^2$ = sum of squares of the sample Y values.

M1.6.9 Find $\sum_{i=1}^{30}(y_i - \overline{y})^2 = SSY$.

M1.6.10 Compute $y_1/y_{13} + y_4 * y_6$.

1.7 Functional Notation

There are no calculations in this section that require Minitab.

1.8 Matrices and Vectors

While Minitab is primarily a program for statistical analysis of data, it can also be used for performing matrix arithmetic. In this manual we describe some of the Minitab commands for matrix calculations that can be used for regression analysis.

Matrix Calculations

First we show you how to create matrices in Minitab and save them in a file. Then we discuss some Minitab commands for various matrix operations.

Consider the two 4×4 matrices $\boldsymbol{A}$ and $\boldsymbol{B}$

$$\boldsymbol{A} = \begin{bmatrix} 12 & 32 & 31 & 27 \\ 38 & 54 & 19 & 10 \\ 65 & 76 & 23 & 24 \\ 24 & 12 & 26 & 52 \end{bmatrix} \qquad \boldsymbol{B} = \begin{bmatrix} 24 & 17 & 27 & 32 \\ 39 & 13 & 15 & 37 \\ 16 & 42 & 26 & 33 \\ 36 & 37 & 23 & 41 \end{bmatrix}$$

We enter (via the keyboard) these two matrices and label them `m1` and `m2`, respectively. In Minitab a matrix can be designated as `M1`, `M2`, ..., or as `m1`, `m2`, ..., etc. Minitab does not distinguish between upper case and lower case for matrix names. The command for entering (via the keyboard) a matrix with `k1` rows and `k2` columns and labeling it `m1` is

COMMAND TO READ A MATRIX INTO THE WORKSHEET VIA THE KEYBOARD

```
MTB > read k1 k2 m1
```

To illustrate, we read the 4×4 matrix $\boldsymbol{A}$ into the worksheet via the keyboard, and label it `m1` using the following command (after each row is entered press `Enter`).

COMMAND TO READ A MATRIX INTO THE WORKSHEET VIA THE KEYBOARD

```
MTB > read 4 4 m1
DATA> 12 32 31 27
DATA> 38 54 19 10
DATA> 65 76 23 24
DATA> 24 12 26 52
```

Minitab responds with:

```
---------------------------------------------------------------------

      4 ROWS READ
---------------------------------------------------------------------
```

Next we read the 4×4 matrix B into the worksheet via the keyboard and label it m2.

COMMAND TO READ A MATRIX INTO A COMPUTER VIA THE KEYBOARD

```
MTB > read 4 4 m2
DATA> 24 17 27 32
DATA> 39 13 15 37
DATA> 16 42 26 33
DATA> 36 37 23 41
```

Minitab responds with:

```
---------------------------------------------------------------------

      4 ROWS READ
---------------------------------------------------------------------
```

As with everything that is entered via the keyboard, you should check these matrices. They may be viewed using the `print` command.

COMMAND TO PRINT MATRICES

```
MTB > print m1 m2
```

Minitab responds with:

```
--------------------------------------------------------------------

 MATRIX M1

  12    32    31    27
  38    54    19    10
  65    76    23    24
  24    12    26    52

 MATRIX M2

  24    17    27    32
  39    13    15    37
  16    42    26    33
  36    37    23    41
--------------------------------------------------------------------
```

We can save the matrices m1 and m2 along with whatever else is part of the worksheet with the save command and the portable subcommand as described below. The file name, including the path, must be supplied by you. The file extension mtp (or MTP) need not be specified explicitly.

COMMAND AND SUBCOMMAND TO SAVE MATRICES IN A PORTABLE WORKSHEET FILE

```
MTB > save 'filename';
SUBC> portable.
```

If no path is specified with the filename in the command above, the file will be stored in the current working directory.

Now we show how to perform some basic matrix calculations. To add the two matrices A and B, i.e., the matrices m1 and m2, and save the result in the matrix m3 we use the following command.

COMMAND TO ADD TWO MATRICES

```
MTB > add m1 m2 m3
```

The result may be viewed by using the command `print m3`.

Minitab responds with:

```
----------------------------------------------------------------------

MATRIX M3

   36     49     58     59
   77     67     34     47
   81    118     49     57
   60     49     49     93
----------------------------------------------------------------------
```

Next we subtract `m1` from `m2` and put the result in `m4`. The command is

COMMAND TO SUBTRACT ONE MATRIX FROM ANOTHER

```
MTB > subtract m1 m2 m4
```

The result is (by using the command `print m4`)

```
MATRIX M4

  12  -15   -4    5
   1  -41   -4   27
 -49  -34    3    9
  12   25   -3  -11
```

So `m4` = `m2` − `m1`. Remember, you must keep track of the matrices yourself; i.e., you must keep a record of which matrix is in `m1`, which matrix is in `m2`, etc. There is no facility in Minitab for naming matrices.

Next we illustrate how to compute the product of two matrices. To multiply `m1` by `m2`, with `m1` on the left and `m2` on the right, and put the result in `m5`, the command is

COMMAND TO MULTIPLY TWO MATRICES

```
MTB > multiply m1 m2 m5
```

The result is (by using the command `print m5`)

```
MATRIX M5

3004   2921   2231   3698
3682   2516   2560   4251
5756   3947   4045   6635
3332   3580   2700   4202
```

Thus `m5` = `m1` times `m2`.

Next we show how to compute the transpose of a matrix. To compute the transpose of `m1` and put the result in `m6`, the command is

COMMAND TO TRANSPOSE A MATRIX

```
MTB > transpose m1 m6
```

The result is (by using the command `print m6`)

```
MATRIX M6

12   38   65   24
32   54   76   12
31   19   23   26
27   10   24   52
```

To compute the inverse of the matrix `m2` and put the result in `m7`, the command is

COMMAND TO INVERT A MATRIX

```
MTB > invert m2 m7
```

The result is (by using the command `print m7`)

```
MATRIX M7

 0.050068 -0.121315 -0.153764  0.194163
-0.015765 -0.052738 -0.021490  0.077193
 0.112490 -0.124162 -0.097838  0.102999
-0.092840  0.223765  0.209290 -0.273536
```

It is instructive to multiply `m2` by its inverse in `m7` and verify that the result is a 4 by 4 identity matrix. We do this using the commands

```
multiply m2 m7 m8
print m8
```

and the result is

```
MATRIX M8

1.0000000  0.0000001  0.0000002 -0.0000002
0.0000000  1.0000001  0.0000002 -0.0000002
0.0000001 -0.0000000  1.0000001 -0.0000000
0.0000001  0.0000001  0.0000002  0.9999999
```

Note that `m8` is indeed the 4 by 4 identity matrix except for rounding errors.

Problems

M1.8.1 Problems (a)–(m) refer to the matrix $\boldsymbol{X}$ and the vector $\boldsymbol{y}$ defined below. For each problem give the appropriate Minitab command and the answer if requested.

$$X = \begin{bmatrix} 12 & 28 & 21 \\ 14 & 31 & 46 \\ 20 & 21 & 31 \\ 11 & 19 & 21 \\ 16 & 13 & 34 \\ 39 & 26 & 30 \\ 25 & 37 & 15 \end{bmatrix} \qquad y = \begin{bmatrix} 9 \\ 13 \\ 28 \\ 6 \\ 32 \\ 16 \\ 24 \end{bmatrix}$$

(a) Read the two matrices X and y into the worksheet via the keyboard and denote them by `m11` and `m12`, respectively. Then print `m11` and `m12` and examine the numbers to verify there are no data entry errors.

(b) Compute X^T and X^TX.

(c) Compute X^Ty.

(d) Compute $(X^TX)^{-1}$.

(e) Compute $(X^TX)^{-1}X^Ty$.

(f) Compute y^Ty.

(g) Compute $y^T[I - X(X^TX)^{-1}X^T]y$.

(h) Compute E where $E = I - (\frac{1}{7})J$, where I is the 7 by 7 identity matrix, and J is a 7 by 7 matrix with each element equal to 1.

(i) Show that $EE = E$.

(j) Show that $y^T[(\frac{1}{7})J]y = 7\bar{y}^2$.

(k) Show that $\bar{y} = (\frac{1}{7})\mathbf{1}^Ty$ where $\mathbf{1}$ is a 7 by 1 vector with each element equal to 1.

(l) Show that $y^TEy = \sum_{i=1}^{7}(y_i - \bar{y})^2 = SSY$.

(m) Show that $EJ = 0$.

1.9 Multivariate Gaussian Populations

To examine large data sets it is sometimes useful to plot them so you can study them visually. In this section we discuss Minitab commands that can be used to plot histograms of single columns of data. We also explain the command to plot two-variable data. For illustrations, we use the data in the two files **bivgauss.mtp** and **bivngaus.mtp** on the data disk.

Histograms

The Minitab command to display a histogram of the data in a given column, say c1, is given below.

HISTOGRAM COMMAND

```
MTB > histogram c1
```

Of course any column can be used in place of column c1 as long as it contains data you want to use. We illustrate the histogram command using the data in the file **bivgauss.mtp**. You should first retrieve this file and examine its contents. The data in this file are in columns c1 and c2 of the worksheet. To obtain a histogram of the X_1 data in column c1 we use the command histogram c1.

Minitab responds with:

```
-----------------------------------------------------------------------

Histogram of X1   N = 1000
Each * represents 5 obs.

Midpoint   Count
     -10       3  *
      -8       8  **
      -6      19  ****
      -4      65  *************
      -2     118  ************************
       0     163  *********************************
       2     224  *********************************************
       4     177  ************************************
       6     128  **************************
       8      56  ************
      10      31  *******
      12       6  **
      14       2  *
-----------------------------------------------------------------------
```

several subcommands that allow you to pick class midpoints, class widths, etc., can be used with the histogram command. You should consult the *Minitab Reference Manual* for details.

Scatter Plots

To plot the data in one column, say c1, against the data in another column, say c2, use the command

PLOT COMMAND

```
MTB > plot c1 c2
```

In the above command the vertical axis is used for c1 and the horizontal axis is used for c2. Of course any two columns of data of the same length can be used in place of c1 and c2. We illustrate the `plot` command using the data in the file **bivngaus.mtp**. To plot c1 against c2 the command is `plot c1 c2`, and we get the following.

Minitab responds with:

```
------------------------------------------------------------------

         -
X1       -            *
         -                    * ***
         -                  **232332
      10+           *     242 3*23 337*4*
         -         3 3 2223*557244848++3
         -    *      ***444+4+8+5++++++2
         -       *2*42*24*437++++++++++4
         -                            +++++++37++73*522* * **    *
       0+                             ++++++7+68564632**4 *    *
         -                            4+95+97263547 22**    2
         -                            42383325422*422* *    *
         -                             * *4  4    *    *
         -                              *   2    *    *
     -10+                             * 2
         -
         -
          ------+---------+---------+---------+---------+---------+X2
              -30       -15         0        15        30        45
------------------------------------------------------------------
```

You will notice that this plot looks similar to the plot in Figure 1.9.6, but this plot is not as detailed as the one there. To obtain high resolution plots you can use the command `gplot c1 c2` (the g in gplot stands for graphics), but you may have to specify the graphics device you wish to use. To learn more about this and other commands and subcommands for plotting in Minitab, consult the *Minitab Reference Manual*.

Problems

Give the Minitab commands required to answer each problem.

M1.9.1 Retrieve the data in the file **bivgauss.mtp**.

M1.9.2 In Problem **M1.9.1** examine what the file contains.

M1.9.3 For the data in the file **bivgauss.mtp**, obtain a histogram of the X_1 data in column c1.

M1.9.4 For the data in the file **bivgauss.mtp**, obtain a histogram of the X_2 data in column c2.

M1.9.5 For the data in the file **bivgauss.mtp**, plot c1 against c2.

M1.9.6 For the data in the file **bivgauss.mtp**, plot c2 against c1.

Chapter 2

Regression and Prediction

2.1 Overview

There are no calculations in this section that require Minitab.

2.2 Prediction

There are no calculations in this section that require Minitab.

2.3 Regression Analysis

In Chapter 1 of this manual we introduced some basic Minitab commands and illustrated their use. Our discussion there was mainly about a one-variable data set, but in this section we introduce some Minitab commands for data sets that contain more than one variable. To illustrate various Minitab commands we consider Example 2.3.1 and Task 2.3.1, both of which use the car data in Table D-1 in Appendix D. These data are also stored in the file **car.mtp** on the data disk. First we retrieve this file and use the `info` command to see what it contains.

Minitab responds with:

```
------------------------------------------------------------------------

COLUMN     NAME       COUNT
C1         carno       1242
C2         mtcost      1242
C3         price       1242
C4         miles       1242

CONSTANTS USED: NONE
------------------------------------------------------------------------
```

We observe that there are four columns (COLUMN) of data, c1, c2, c3, and c4, and that each column consists of 1242 (shown under COUNT) numbers. The names of the variables associated with each column of numbers are also given in the column labeled NAME.

At this stage, you should display the data on the screen and examine them. This is done using the command

```
print c1 c2 c3 c4
```

Minitab responds with:

```
------------------------------------------------------------------------

 ROW  carno  mtcost   price   miles

   1      1     551   36400   12400
   2      2     661   15200   15400
   3      3     679   14100   16000
   4      4     561   22500   12100
   5      5     497   20600   11200

 ....
 ....
```

```
1238   1238    381  23600   8000
1239   1239    464  30700   7300
1240   1240    563  21100  12000
1241   1241    602  22300  14000
1242   1242    582  14600  13100
---------------------------------------------------------------------
```

To save space, we have reproduced only the first five lines and the last five lines of data here.

It is often desirable to compute various summary statistics for each column (i.e., for each variable). These can be obtained using the command

```
describe c2 c3 c4
```

Minitab responds with:

```
---------------------------------------------------------------------

              N     MEAN   MEDIAN   TRMEAN    STDEV   SEMEAN
mtcost     1242   526.14   502.00   520.02   105.97     3.01
price      1242    19648    19400    19497     5838      166
miles      1242  11114.5  11100.0  11154.1   3084.4     87.5

            MIN      MAX       Q1       Q3
mtcost   352.00   925.00   445.00   592.00
price      7200    38300    15200    23625
miles    1600.0  18500.0   9000.0  13325.0

---------------------------------------------------------------------
```

From this output you can obtain several statistics for each of the variables of interest. Note that we did not request Minitab to print summary statistics for the variable carno since these are not very meaningful.

Example M2.3.1

Here we illustrate Minitab commands that can be used to do the computations required in Example 2.3.1. First we obtain the data in Table 2.3.1, which is only a subset of the data in the file **car.mtp** on the data disk. We want the maintenance costs of cars that were driven 14,000 miles the first year. The following Minitab command and subcommand extract the data values corresponding to these cars. You should first retrieve the file **car.mtp**.

COMMAND FOR OBTAINING A SUBPOPULATION

```
MTB > copy c1 c2 c3 c4 c11 c12 c13 c14;
SUBC> use c4=14000.
```

The main command `copy c1 c2 c3 c4 c11 c12 c13 c14` asks Minitab to copy the numbers in columns c1, c2, c3, and c4, respectively, into columns c11, c12, c13, and c14; the subcommand `use c4=14000` instructs Minitab to copy only those data points for which c4 equals 14,000. For these points, the elements in column c1 are copied into column c11, the elements in column c2 are copied into column c12, the elements in column c3 are copied into column c13, and the elements in column c4 are copied into column c14. Now we print columns c11, c12, c13, and c14 and examine them using the command

```
print c11 c12 c13 c14
```

Minitab responds with:

```
--------------------------------------------------------------------

 ROW    C11    C12    C13    C14

   1     78    656  16100  14000
   2    209    633  18400  14000
   3    382    637  12900  14000
   4    402    612  22000  14000
   5    626    624  21900  14000
   6    641    620  18100  14000
   7    777    605  17000  14000
   8    888    607  13300  14000
   9    891    654  25600  14000
  10    928    620  17300  14000
  11   1029    622  16500  14000
  12   1030    645   9500  14000
  13   1040    567  15300  14000
  14   1093    596  23700  14000
```

```
    15    1199    639    13600    14000
    16    1241    602    22300    14000
--------------------------------------------------------------------------
```

Notice that this subpopulation contains 16 cars, each of which was driven 14,000 miles during the first year after purchase. The car numbers of cars in this group are in column c11, the maintenance costs are in column c12, prices are in c13, and miles driven are in c14. Columns c12 and c14 make up Table 2.3.1.

The mean and the standard deviation of the first-year maintenance costs of cars in the subpopulation of cars that were driven 14,000 miles during the first year, are obtained using the command

```
describe c12
```

Minitab responds with:

```
--------------------------------------------------------------------------

                  N      MEAN    MEDIAN    TRMEAN     STDEV    SEMEAN
C12              16    621.19    621.00    622.57     23.18      5.80

                MIN       MAX        Q1        Q3
C12          567.00    656.00    605.50    638.50
--------------------------------------------------------------------------
```

Thus $\mu_Y(14,000) = \$621.19$ and $\sigma_Y(14,000) = \$23.18$, which are of course the same as the values obtained in Example 2.3.1.

Next we demonstrate some Minitab commands that can be used to obtain some of the quantities in Task 2.3.1.

Task M2.3.1

Once again we use the data given in Table D-1 in Appendix D and in the file **car.mtp** on the data disk. You should retrieve this file first. In parts (1a) and (1b) of this task, we need a histogram as well as the mean and the standard deviation of the first-year maintenance costs of all cars in the population. These can be obtained with the Minitab commands that we have discussed earlier.

In part (2a), suppose we want the Y value of car number 354. We can obtain this value from observation 354 in Table D-1, or we can obtain it by using the Minitab commands

```
let k1=c2(354)
print k1
```

The command `let k1=c2(354)` puts the 354th item of column c2 into k1. The command `print k1` prints the value of k1.

Minitab responds with:

```
-------------------------------------------------------------------------

K1         483.000
-------------------------------------------------------------------------
```

Thus, the first-year maintenance cost for car number 354 is $483.

For part (3) of Task 2.3.1 we use the command `plot c2 c4`, which results in a plot of the first-year maintenence costs against miles driven.

Problems

Problems **M2.3.1–M2.3.15** refer to the population data given in Table D-1 in Appendix D, which are also stored in the ASCII file **car.dat** and the Minitab worksheet file **car.mtp** on the data disk. For each problem, give the appropriate Minitab command and the answer.

M2.3.1 Read the data values from the file **car.dat** and put them in columns c1, c2, c3, and c4. Name the variables as follows: give c1 the name I; give c2 the name Y; give c3 the name $X1$; give c4 the name $X2$.

M2.3.2 Compute the mean and the standard deviation of Y and $X1$.

M2.3.3 What are the minimum and maximum values of the $X1$ data? Of the $X2$ data?

M2.3.4 Print the Y data.

M2.3.5 Plot a histogram of the Y data.

M2.3.6 Plot a histogram of the $X1$; also of the $X2$ data.

M2.3.7 Compute SSY.

M2.3.8 Compute $\sum_{i=1}^{1242} Y_i^2$.

M2.3.9 Find the standard deviation of Y; of $X1$; of $X2$.

M2.3.10 What are the minimum and the maximum of the Y values?

M2.3.11 Rename C3 as X.

M2.3.12 Plot the values of Y against X.

M2.3.13 What was the first-year maintenance cost for car number 792?

M2.3.14 **(a)** How many cars in the study population sold for $12,500?
(b) Which ones are they (give their item numbers)?
(c) Explicitly list the first-year maintenance costs associated with these cars.
(d) Calculate the mean and the standard deviation of the first-year maintenance costs for these cars.

M2.3.15 Predict the maintenance cost of a car randomly chosen from the study population, if you know this car sold for $9,600.

Chapter 3

Straight Line Regression

3.1 Overview

In this chapter we show how Minitab can be used to compute many of the quantities needed in straight line regression. There are a few quantities for which no direct commands are available in the present version of Minitab, and for these we have written programs that we sometimes refer to as **macros**. These programs are supplied to you on the data disk that accompanies this manual.

As usual, in each section of the manual we discuss Minitab computing instructions pertaining to corresponding sections of the textbook.

3.2 An Example of Straight Line Regression

All of the computations required in Section 3.2, such as `retrieve` , `info` , `plot` , `histogram` , `describe` , etc., have been discussed in Chapters 1 and 2 of this manual. You should refer those chapters for information about these commands.

3.3 Straight Line Regression Model – Assumptions (A) and (B)

There are no calculations in this section that require Minitab.

3.4 Point Estimation

In this section we explain how Minitab can be used to compute point estimates of parameters in straight line regression. The Minitab command to compute many of the quantities needed in regression is `regress` . For straight line regression this command will ask Minitab to compute the least squares estimates of β_0, β_1, and σ, as well as many other useful quantities.

In its simplest form the `regress` command consists of **three parts** – the first part is the column containing the response variable Y (say Y is in column `c1`); the second part is the number of predictor variables in the regression function (which is always 1 for straight line regression); the third part is the column containing the predictor variable X (say X is in column `c2`). Then the `regress` command, in its simplest form, for straight line regression is

REGRESSION COMMAND

```
MTB > regress c1 1 c2
```

We will explain the use of this command by referring to Task 3.4.1, where an investigator is studying the relationship of Y, the weight of crystals, to X, the number of hours they are allowed to grow. The data are given in Table 3.4.2 and are also stored in the file **crystal.mtp** on the data disk. As usual, prior to carrying out regression calculations, you should retrieve the file **crystal.mtp** and examine its contents. Then you should plot the data and examine them for abnormalities or obvious violations of assumptions. We will assume throughout that you will automatically carry out these preliminary steps.

For the crystal data, recall that the response variable Y is in column `c1` and the predictor variable X is in column `c2`. So the `regress` command for obtaining point estimates of β_0, β_1, and σ is

```
MTB > regress c1 1 c2
```

Minitab responds with:

```
-----------------------------------------------------------------------
The regression equation is

weight =  0.001 + 0.503 time                                   (M3.4.1)

Predictor       Coef       Stdev    t-ratio        p
Constant      0.0014      0.5994       0.00    0.998           (M3.4.2)
time         0.50343     0.03520      14.30    0.000           (M3.4.3)

s = 1.062        R-sq = 94.5%     R-sq(adj) = 94.0%            (M3.4.4)
-----------------------------------------------------------------------
```

To save space, we list only those parts of the computer output that are relevant to the problem being discussed, so you should not be surprised if the output you obtain is more than what we list. The equation numbers shown as part of the Minitab output above were added by us for convenience of reference and were not printed by Minitab.

Note that the estimated regression function, $\hat{\mu}_Y(x) = \hat{\beta}_0 + \hat{\beta}_1 x$, is given in (M3.4.1) from which we can obtain the values of $\hat{\beta}_0$ and $\hat{\beta}_1$. The estimated values of β_0 and β_1 are also given in (M3.4.2) and (M3.4.3), respectively, under the column labeled `Coef`. The quantity labeled `s` in (M3.4.4) is the estimated value of σ and it is equal to 1.062. You can see that the values of $\hat{\beta}_0$, $\hat{\beta}_1$, and $\hat{\sigma}$ are the same as those obtained in Task 3.4.1. Other quantities in the output of the `regress` command will be discussed later.

In its complete form, the `regress` command is capable of processing additional optional arguments as well as optional subcommands. We will discuss these later, but if you are curious, you can consult the *Minitab Reference Manual* for more details.

Brief

The quantity of output that is obtained when Minitab executes the `regress` command can be controlled by the user with the help of the `brief` command. Before you give the `regress` command, you can give any one of the four commands, `brief 0`, `brief 1`, `brief 2`, or `brief 3`. The `brief 0` command suppresses all output and is useful for programming special purpose macros. The `brief 1`, `brief 2`, and `brief 3` commands ask Minitab to output increasing amounts of information, `brief 3` resulting in the most detailed output. If you don't issue a `brief` command, then Minitab uses `brief 2` by default. A `brief` command stays in effect until you give another `brief` command or until you exit from Minitab (which you can do by giving the command `stop`). You should try each of the `brief` commands with the `regress` command and observe the resulting quantity of output.

Problems

For each problem give the appropriate Minitab command and the answer if required.

M3.4.1 For the data in Problem 3.2.1 given in Table 3.2.3 and stored in the file **table323.dat** on the data disk, read the data into columns `C1` and `C2` and print them.

M3.4.2 For Problem **M3.4.1**, use the `regress` command and compute estimates of β_0, β_1, $\mu_Y(x)$, and σ.

M3.4.3 Use the command `brief 1` and then work Problem **M3.4.2**. Has the amount of output changed? Use the command `brief 2` and work Problem **M3.4.2** again. Has the amount of output changed? Use the command `brief 3` and then work Problem **M3.4.2**. Has the amount of output changed? Use the command `brief 0` and then work Problem **M3.4.2**. Has the amount of output changed?

M3.4.4 Repeat Problems **M3.4.1** and **M3.4.2** for the data of Problem 3.2.5, which are given in Table 3.2.4 and are stored in the file **table324.dat**.

M3.4.5 Consider the data in Table 3.4.3, which are stored in the file **arsenic.dat** on the data disk. Read the Y data into column `c13` and the X data into column `c21`. Print the data. Save them in a Minitab worksheet file named **t345.mtp** and store this file in the subdirectory `c:`.

M3.4.6 Retrieve the data from the file **t345.mtp** in the subdirectory `c:` in Problem **M3.4.5**.

M3.4.7 In Problem **M3.4.6** compute the estimates of β_0, β_1, $\mu_Y(x)$, and σ.

3.5 Checking Assumptions

In this section we show how Minitab commands can be used to perform many of the calculations for residual analyses discussed in Section 3.5. You should be familiar with the material in that section before attempting to learn the Minitab commands discussed here. Recall that residual analyses are useful for examining the validity of assumptions **(A)** and/or assumptions **(B)**.

We will demonstrate the commands to compute the following:

(1) Fitted values (fits), $\hat{\mu}_Y(x)$

(2) Residuals, $\hat{e}_i$

(3) Hat values, $h_{i,i}$

(4) Standardized residuals, r_i

(5) Gaussian scores (nscores), $z_i^{(n)}$

All of the above quantities except the Gaussian scores are obtained by using an extension of the simple `regress` command and suitable subcommands. To illustrate these commands and subcommands we use the crystal data given in Table 3.4.2, which are stored in the file **crystal.mtp** on the data disk. We exhibit the Minitab commands that are used to obtain the results in Example 3.5.1. We do not explain those Minitab commands that have previously been discussed, but we use them when necessary. You should retrieve the data from the file **crystal.mtp**, print them, and check them before proceeding.

The extended `regress` command and the relevant subcommands are

REGRESS COMMAND AND DIAGNOSTIC SUBCOMMANDS

```
MTB > regress c1 1 c2 c3 c4;
SUBC> residuals c5;
SUBC> hi c6.
```

The extended regress command `regress c1 1 c2 c3 c4;` consists of five components. It instructs Minitab to do the following: regress Y (weight), which is stored in `c1`, using 1 predictor variable X (time), which is stored in `c2`, calculate the **standardized residuals** and put them in `c3`, and finally calculate the **fits** and put them in `c4`. This command ends with a semicolon, which tells Minitab that a subcommand is to follow. The first subcommand is `residuals c5;` , and it instructs Minitab to compute the **residuals** and put them in column `c5`. Note that this subcommand ends with a semicolon also, which tells Minitab to expect another subcommand. The next subcommand is `hi c6.` , which asks Minitab to compute the hat values and put them in column `c6`. This subcommand ends with a period, which tells Minitab to expect no more subcommands.

There are many other Minitab subcommands that can be used with the `regress` command to perform additional calculations, but here we discuss only those subcommands that we need for residual analysis. Consult the *Minitab Reference Manual* for a complete list of subcommands that can follow the `regress` command.

Note that the last two columns (`c3` and `c4` here) specified in the extended version of the `regress` command are optional columns, and when specified, they are reserved for storing the standardized residuals and fitted values (fits) *in this order*. Thus the first of the optional columns specified (`c3` here) is reserved for storing standardized residuals, and the second of the optional columns (`c4` here) is reserved for the fitted values. If only one optional column is specified, then this column will be used to store the standardized residuals. For example, the command

```
regress c1 1 c2 c7 c4
```

will instruct Minitab to store the standardized residuals in `c7` and the fitted values in `c4`, whereas the command

```
regress c1 1 c2 c8
```

will ask Minitab to store the standardized residuals in `c8` (the fitted values will not be stored in any of the columns in this case).

The Minitab response to the command and subcommands in the preceding box is (as usual, we print only the portion of the Minitab output that is needed for our discussion).

```
--------------------------------------------------------------------------

The regression equation is
weight = 0.001 + 0.503 time

Predictor        Coef        Stdev    t-ratio        p
Constant       0.0014       0.5994       0.00    0.998
time          0.50343      0.03520      14.30    0.000

s = 1.062        R-sq = 94.5%      R-sq(adj) = 94.0%

--------------------------------------------------------------------------
```

You should note that the standardized residuals, the fits, the residuals, and the hat values that were computed are not printed automatically. You must give the appropriate command to have them printed, but first, as we have stated before, it is important to name the columns of numbers that have been computed so they can be easily identified. We give column c3 the name `stdresid` and column c4 the name `fits`. We give column c5 the name `residual`, and the column c6 the name `hatvals`. The commands are

COMMAND TO NAME COLUMNS

```
MTB > name c3 'stdresid'
MTB > name c4 'fits'
MTB > name c5 'residual'
MTB > name c6 'hatvals'
```

The quantiites Y (weight), X (time), standardized residuals, fitted values, residuals, and hat values can be printed and plotted so they can be examined to help assess whether the assumptions for regression are satisfied. You should print these quantities and compare them with those in Table 3.5.1. Note that if you use the `regress` command after setting `brief` equal to 3, Minitab will automatically print fits, residuals, and standardized residuals as part of the output.

Next we obtain the Gaussian scores (nscores) of the standardized residuals, put them in column c7, give them the name `nscores`, and plot them, with the following commands.

COMMANDS FOR NSCORES and RANKIT PLOT

```
MTB > nscore c3 c7
MTB > name c7 'nscores'
MTB > plot c3 c7
```

The first command,

`nscore c3 c7`

computes the Gaussian scores, i.e., <u>nscores</u>, of the standardized residuals r_i that are in column c3, and puts them in column c7. The second command

`name c7 'nscores'`

gives column c7 the name 'nscores' so they can be easily identified. The third command

`plot c3 c7`

instructs Minitab to plot the standardized residuals against the nscores.

Minitab responds with:

```
---------------------------------------------------------------------------

        -                                                      *
     1.2+                                               *
        -                                       *   *
stdresid-                                    *
        -
        -                                  *
     0.0+                          *  * *
        -                        *
        -
        -
        -                 *   *
    -1.2+
        -             *
        -
        -      *
        -
    -2.4+
           --------+---------+---------+---------+---------+--------nscores
               -1.40     -0.70      0.00      0.70      1.40

---------------------------------------------------------------------------
```

This plot is also given in Figure 3.5.21. The points appear to lie (approximately) on a straight line of unit slope through the origin.

Problems

Problems **M3.5.1** through **M3.5.6** refer to Example 3.5.2. For each problem, give the appropriate Minitab commmands and the answer when applicable.

M3.5.1 Retrieve the data from the file **car20.mtp** on the data disk; print and examine the contents of this file.

M3.5.2 Plot Y (maintenance cost) against X (miles driven).

M3.5.3 Regress Y (maintenance cost) on X (miles driven), obtain the standardized residuals r_i, the residuals $\hat{e}_i$, and the fitted values $\hat{\mu}_Y(x_i)$. Put these quantities in suitable columns, and name these columns.

M3.5.4 Obtain point estimates of β_1, β_0, $\mu_Y(x)$, and σ.

M3.5.5 Find $\hat{\mu}_Y(9400)$.

M3.5.6 Use the `describe` command to compute $\hat{\mu}_Y$, $\hat{\mu}_X$, $\hat{\sigma}_Y$, and $\hat{\sigma}_X$.

3.6 Confidence Intervals

The output from the `regr` command gives the quantities $\hat{\beta}_0$, $\hat{\beta}_1$, and the standard errors of these quantities. You can use these in (3.6.1) and compute confidence intervals for β_0 and β_1. The output from this command also gives $\hat{\sigma}$, and you can use this in (3.6.8) to compute confidence intervals for σ. Minitab does provide a subcommand

```
predict x
```

with the `regress` command that will directly output 95% confidence intervals for $\mu_Y(x)$ and 95% prediction intervals for $Y(x)$ for a user-specified value x. However, there are no built-in commands or subcommands in Minitab to compute $1-\alpha$ confidence intervals for the following:

(a) σ_Y

(b) σ

(c) $\mu_Y(x)$ (except 95% intervals)

(d) β_0

(e) β_1

(f) $a_0\beta_0 + a_1\beta_1$ for general values of a_0 and a_1

(g) $Y(x)$ (except 95% intervals)

Actually a 95% confidence interval for β_0 can be obtained with the Minitab subcommand `predict 0` of the `regress` command since this computes a 95% confidence interval for $\mu_Y(0)$, which is β_0.

We have written three macros – named **sgmaconf**, **confint**, and **predy** – which, collectively, compute confidence intervals for all of the quantities in items **(a)**–**(g)** above. These macros are provided on the data disk. In this section we demonstrate the use of the `predict` subcommand and also the use of the above macros.

The PREDICT Subcommand

The `predict` subcommand of the `regress` command computes 95% two-sided confidence intervals for $\mu_Y(x)$, $Y(x)$, and β_0. If the Y data are in column `c1` and the X data are in column `c2`, the command and subcommand are (`x` in the subcommand is a number specified by you)

COMMAND FOR CONFIDENCE INTERVALS FOR $Y(x)$ and $\mu_Y(x)$

```
MTB > regress c1 1 c2;
SUBC> predict x.
```

To illustrate, we use the data of Task 3.6.2, which are in Table 3.6.2 and are also stored in the file **agebp.mtp** on the data disk. You should retrieve this file and examine its contents. The response variable Y is the average systolic blood pressure taken at 8 am over a two week period, and the predictor variable X is the age of the individual in years. The values of Y are stored in column c1 and the values of X are stored in c2. Suppose we want a 95% two-sided confidence interval and prediction interval for $\mu_Y(x)$ and $Y(x)$, respectively, for $x = 60$. The commmand and the subcommand are

COMMAND FOR CONFIDENCE INTERVALS
FOR $Y(60)$ and $\mu_Y(60)$

```
MTB > regress c1 1 c2;
SUBC> predict 60.
```

The Minitab response is (only the relevant portion of the output is displayed)

```
---------------------------------------------------------------------------
The regression equation is
bp = 66.8 + 1.61 age                                                (M3.6.1)

Predictor        Coef       Stdev    t-ratio        p
Constant       66.808       2.200      30.37    0.000               (M3.6.2)
age           1.60853     0.04720      34.08    0.000               (M3.6.3)

s = 2.836       R-sq = 98.1%     R-sq(adj) = 98.1%                  (M3.6.4)

     Fit  Stdev.Fit         95% C.I.            95% P.I.
 163.320      0.916   (161.420,165.220)  (157.139,169.501)          (M3.6.5)
---------------------------------------------------------------------------
```

The following information that relates to confidence intervals is obtained from the Minitab output above. You should check these results with those obtained in Task 3.6.2.

(1) The estimated regression equation is $\hat{\mu}_Y(x) = 66.8 + 1.61x$, which is in (M3.6.1).

(2) $\hat{\beta}_0 = 66.808$ and $\hat{\beta}_1 = 1.60853$ are given in (M3.6.2) and (M3.6.3), respectively, under coef.

(3) The standard errors of $\hat{\beta}_0$ and $\hat{\beta}_1$ are also given in (M3.6.2) and (M3.6.3) under Stdev. The values are $SE(\hat{\beta}_0) = 2.200$ and $SE(\hat{\beta}_1) = 0.04720$.

(4) In (M3.6.4) s $= \hat{\sigma} = 2.836$.

(5) The estimate of $\mu_Y(60)$ is 163.32, which is given in (M3.6.5) under `fit`. Recall that

$$\hat{Y}(x) = \hat{\mu}_Y(x) \text{ so } \hat{Y}(60) = \hat{\mu}_Y(60) = 163.32$$

(6) The quantity in (M3.6.5) under `Stdev.Fit` is

$$SE(\hat{\mu}_Y(60)) = 0.916$$

(7) The 95% confidence interval for $\mu_Y(60)$ is under `95% C.I.` in (M3.6.5) and is

$$161.420 \leq \mu_Y(60) \leq 165.220$$

(8) The 95% prediction interval for $Y(60)$ is under `95% P.I.` in (M3.6.5) and is

$$157.139 \leq Y(60) \leq 169.501$$

If we want 95% confidence intervals for $\mu_Y(x)$ and/or $Y(x)$ for several different values of x, say $x = 40, 60, 80$, etc., the command and subcommands are

COMMAND FOR SEVERAL CONFIDENCE INTERVALS FOR $Y(x)$ and $\mu_Y(x)$

```
MTB > regress c1 1 c2;
SUBC> predict 40;
SUBC> predict 60;
SUBC> predict 80.
```

Notice where the semicolons and the period are placed.

Macros for Computing Confidence Intervals

We now discuss the use of the macros **sgmaconf**, **confint**, and **predy**, for computing confidence intervals for quantities for which there are no direct Minitab commands or subcommands. These macros are sets of Minitab commands that are stored in various files in the subdirectory called `macro` on the data disk, and are executed by Minitab when it is instructed to do so. The functions of these macros are as follows:

- **sgmaconf**: This macro can be used to compute confidence intervals for σ_Y and σ.
- **confint**: This macro can be used to compute confidence intervals for $a_0\beta_0 + a_1\beta_1$. In particular, this macro can be used to compute confidence intervals for $\mu_Y(x)$, β_0, and β_1 by choosing appropriate values for a_0 and a_1.
- **predy**: This macro can be used to compute prediction intervals for $Y(x)$.

Next we explain how to use these three macros.

Confidence Intervals for σ and σ_Y

The macro **sgmaconf** may be used to compute confidence intervals for σ (respectively, σ_Y). The Minitab commands for this macro are stored in the file **sgmaconf.mtb** in the subdirectory `macro` on the data disk. Note that the filename has the extension **mtb** so that Minitab can automatically recognize this as a **Minitab macro** file. During the execution of this macro you will be asked to input the value of $\hat{\sigma}$ (respectively, $\hat{\sigma}_Y$) and the associated degrees of freedom $n - 2$ (respectively, $n - 1$). The prompt `DATA >` will be displayed when the macro asks you to input a quantity. Following this prompt, you type the value of the quantity requested and press the `Enter` key.

To illustrate the use of this macro, we compute a 90% two-sided confidence interval for σ in the blood pressure problem of Task 3.6.2. From that task, or from (M3.6.4), we get $\hat{\sigma} = 2.836$. Of course the degrees of freedom are $df = n - 2 = 22$. To start the macro the command is

MACRO COMMAND FOR CONFIDENCE INTERVALS FOR σ

```
MTB > execute 'b:\macro\sgmaconf'
```

Minitab responds with:

```
------------------------------------------------------------------

This program will compute a two-sided confidence interval for
a standard deviation of a one-variable population {Y} or for a
standard deviation in a regression problem. You will be asked
to input the estimate of the standard deviation. You must also
```

```
  input the associated degrees of freedom and the confidence
  coefficient.

*********************************************************************

  Input the estimate of sigma_Y or sigma, whichever applies
DATA> 2.836

  Input the associated degrees of freedom
DATA> 22

  Input the confidence coefficient 1-alpha
DATA> .90
=====================================================================

  For a two-sided  90.0% confidence interval for sigma

  the lower confidence bound is        2.28382     and

  the upper confidence bound is        3.78700
---------------------------------------------------------------------
```

Thus the confidence statement is

$$C[2.28382 \leq \sigma \leq 3.78700] = 0.90$$

Next we compute a 90% two-sided confidence interval for σ_Y for the blood pressure data in Task 3.6.2. We need $\hat{\sigma}_Y$, which may be obtained from Task 3.6.2 or by using the `describe` command. We get $\hat{\sigma}_Y = 20.34$ with the associated degrees of freedom $df = n - 1 = 23$. To start the program, give the command

MACRO COMMAND FOR CONFIDENCE INTERVALS FOR σ_Y

```
MTB > execute 'b:\macro\sgmaconf'
```

Minitab responds with:

```
------------------------------------------------------------------------

  This program will compute a two-sided confidence interval for
  a standard deviation of a one-variable population {Y} or for a
  standard deviation in a regression problem. You will be asked
  to input the estimate of the standard deviation. You must also
  input the associated degrees of freedom and the confidence
  coefficient.

************************************************************************

  Input the estimate of sigma_Y or sigma, whichever applies
DATA> 20.34

  Input the degrees of freedom associated with the estimate
  of sigma
DATA> 23

  Input the confidence coefficient 1-alpha
DATA> .90

========================================================================

  For a two-sided  90.0% confidence interval for sigma

  the lower confidence bound is        16.4480     and

  the upper confidence bound is        26.9610

------------------------------------------------------------------------
```

Thus the confidence statement is

$$C[16.4480 \leq \sigma_Y \leq 26.9610] = 0.90$$

Confidence Interval for $\theta = a_0\beta_0 + a_1\beta_1$

The macro **confint** can be used for computing a confidence interval for the linear combination

$$\theta = a_0\beta_0 + a_1\beta_1$$

in straight line regression. You will be asked to input the column that contains the Y values and the column that contains the X values. You will also be asked to input the values of a_0 and a_1, and the confidence coefficient. Remember, this macro can be used to obtain confidence intervals for the following

(1) β_0, by setting $a_0 = 1$ and $a_1 = 0$.

(2) β_1, by setting $a_0 = 0$ and $a_1 = 1$.

(3) $\mu_Y(x) = \beta_0 + \beta_1 x$, by setting $a_0 = 1$ and $a_1 = x$ for any value x.

(4) $\mu_Y(x_1) - \mu_Y(x_2)$, by setting $a_0 = 0$ and $a_1 = x_1 - x_2$.

the Minitab commands for this macro are stored in the file **confint.mtb** in the subdirectory macro on the data disk. Note that this filename has the extension **mtb** so that Minitab can recognize this file as a **Minitab macro** file. We now illustrate the use of this macro by computing a 95% two-sided confidence interval for β_1 using the age-blood pressure data of Task 3.6.2. Note that in this case $a_0 = 0$ and $a_1 = 1$, so $\theta = a_0\beta_0 + a_1\beta_1 = \beta_1$. You should first retrieve the file **agebp.mtp** and examine its contents. To start the macro the command is

MACRO COMMAND FOR CONFIDENCE INTERVALS FOR $a_0\beta_0 + a_1\beta_1$

```
MTB > execute 'b:\macro\confint'
```

Minitab responds with:

```
------------------------------------------------------------------------

  This macro will compute a confidence interval for theta where
  theta is any linear combination

           a(0) beta(0) + a(1) beta(1)

  of the parameters beta(0) and beta(1).
```

```
  Confidence intervals for beta(0), beta(1), and mu_Y(x) may be
  obtained as special cases by properly choosing a(0) and a(1).

  You are required to input the coefficients a(0) and a(1).

  You are also required to specify the confidence coefficient.

  Leave columns c80-c100 empty for work space.

***********************************************************************

  Input the column number for the column containing Y.
DATA> 1

  Input the column number for the column containing X.
DATA> 2

  Input the constants a(0) and a(1) in the linear function of
  interest in the indicated order.
DATA> 0 1

  Input the confidence coefficient for a two-sided confidence
  interval.
DATA> .95

======================================================================

  The estimated value of theta is     1.608532

  The standard error of this estimate is   0.04720155

  For a two-sided confidence interval for theta with
                            confidence coefficient  0.950

  the lower confidence bound is     1.510641   and

  the upper confidence bound is     1.706423

----------------------------------------------------------------------
```

Thus the point estimate is $\hat{\beta}_1 = 1.608532$ and the standard error of the estimate is $SE(\hat{\beta}_1) = 0.04720155$. The confidence statement is

$$C[1.510641 \le \beta_1 \le 1.706423] = 0.95$$

You should compare these answers with those obtained in Task 3.6.2 to see that they are the same (within rounding error).

Prediction Interval for $Y(x)$

As stated earlier, Minitab has a built in command to compute a 95% two-sided prediction interval for $Y(x)$ for any value x, but it will not compute a prediction interval with confidence coefficient different from 95%. The macro **predy** will compute a prediction interval for $Y(x)$ with any confidence coefficient. The Minitab commands for this macro are in the file **predy.mtb** in the subdirectory macro on the data disk. When using this macro you will be asked to input the column that contains the Y data and the column that contains the X data. You will also be asked to input the value of x and the confidence coefficient $1-\alpha$.

We illustrate the use of this macro by computing a 90% two-sided prediction interval for $Y(38)$ for the age-blood pressure data of Task 3.6.2. These data are stored in the file **agebp.mtp** on the data disk. The Y data are in column c1, and the X data are in column c2. The value of x to use is 38, and $1 - \alpha = 0.90$. First retrieve the file **agebp.mtp** and examine its contents. The command to start the macro is

MACRO COMMAND FOR PREDICTION INTERVALS FOR $Y(x)$

```
MTB > execute 'b:\macro\predy'
```

Minitab responds with:

```
------------------------------------------------------------------------

This program will do calculations necessary to compute a
two-sided prediction interval with confidence coefficient
1-alpha.

**********************************************************************
```

```
   Input the column number that contains the Y data.
DATA> 1


   Input the column number that contains the X data.
DATA> 2


   Input the value of x for which you want a prediction interval
   for Y(x).
DATA> 38


   Input the value of 1-alpha, the confidence coefficient.
DATA> 0.90

=======================================================================

   The point estimate of Y(x) is          127.932

   The standard error of this estimate is          2.91266

   For a two-sided    90% prediction interval for Y(x)
   at the point x =          38.0000

   the lower confidence bound is          122.931     and

   the upper confidence bound is          132.934
-----------------------------------------------------------------------
```

From this we get $\hat{Y}(38) = 127.932$, and the confidence statement is

$$C[122.931 \leq Y(38) \leq 132.934] = 0.90$$

Problems

Problems **M3.6.1–M3.6.12** refer to the arsenic data in Task 3.6.1, which are given in Table 3.6.1 and are also stored in the file **arsenic.mtp**. Exhibit the Minitab commands that can be used to compute quantities needed and give the answers where appropriate.

M3.6.1 Find $\hat{\beta}_0$, $\hat{\beta}_1$, $\hat{\mu}_Y(x)$, $\hat{\sigma}$, $SE(\hat{\beta}_0)$, and $SE(\hat{\beta}_1)$.

M3.6.2 Plot Y against X.

M3.6.3 Compute the residuals $\hat{e}_i$, the fitted values $\hat{\mu}_Y(x_i)$, and the standardized residuals r_i.

M3.6.4 Plot the standardized residuals r_i against each of the following: y_i, x_i, $\hat{\mu}_Y(x_i)$.

M3.6.5 Compute a 90% confidence interval for $\mu_Y(0)$, i.e., for β_0.

M3.6.6 Compute a 90% confidence interval for β_1.

M3.6.7 Obtain $\hat{\mu}_Y(3)$.

M3.6.8 Compute a 95% confidence interval for $\mu_Y(3)$.

M3.6.9 Find $\hat{Y}(3)$.

M3.6.10 Compute a 95% prediction interval for $Y(3)$.

M3.6.11 Find $\hat{\mu}_Y(2)$ and the standard error of $\hat{\mu}_Y(2)$.

M3.6.12 Compute an 85% confidence interval for $\mu_Y(2)$.

3.7 Tests

The only Minitab command that will directly perform tests for regression is the command `regress` , which will compute P-values to test the following

(1) NH: $\beta_0 = 0$ versus AH: $\beta_0 \neq 0$,

(2) NH: $\beta_1 = 0$ versus AH: $\beta_1 \neq 0$, and

(3) NH: $\beta_0 = \beta_1 = 0$ versus AH: at least one of β_0 and β_1 is not zero

For the tests in (1) and (2) above, Minitab prints the P-values along with the point estimates and the corresponding standard errors for $\hat{\beta}_0$ and $\hat{\beta}_1$. For an example refer to (M3.6.2) and (M3.6.3). The P-value for the test in (3) is obtained as part of an analysis of variance table. This will be discussed in Section 3.8.

Since it may be desirable to have more general tests than the ones listed above, we have written a macro named **test**, which is in the file **test.mtb** in the subdirectory called `macro` on the data disk, that will carry out the computations for testing

$$\text{NH: } \theta = q \text{ versus AH: } \theta \neq q$$

or

$$\text{NH: } \theta \leq q \text{ versus AH: } \theta > q$$

or

$$\text{NH: } \theta \geq q \text{ versus AH: } \theta < q$$

where θ is defined by

$$\theta = a_0\beta_0 + a_1\beta_1$$

which is a linear combination of the regression coefficients β_0 and β_1 in a straight line regression model $\mu_Y(x) = \beta_0 + \beta_1 x$. You can use a test about θ to perform tests about β_0, β_1, and $\mu_Y(x)$ by selecting the appropriate values for a_0 and a_1. The procedure for performing these tests is in Box 3.7.4 in the textbook. When you use the macro **test**, you will be asked to input the column number that contains the Y data and the column number that contains the X data. You will also be asked to input the values of a_0, a_1, and q.

To illustrate the use of this macro, we refer to the arsenic data of Task 3.7.1 where an investigator is interested in using the data to help determine whether $\beta_0 = 0$ and $\beta_1 = 1$. In this context we may want to test

$$\text{NH: } \beta_0 = 0 \text{ versus AH: } \beta_0 \neq 0$$

The data are stored in the file **arsenic.mtp** on the data disk. You should first retrieve this file and examine its contents. Note that for this test $a_0 = 1, a_1 = 0$, and $q = 0$. The command to start the macro is

MACRO COMMAND FOR TESTING THE LINEAR COMBINATION $a_0\beta_0 + a_1\beta_1$

```
MTB > execute 'b:\macro\test'
```

Minitab responds with:

```
--------------------------------------------------------------------

  This program will  compute the test statistic t_C for testing a
  linear combination

              theta=a(0) beta(0) + a(1) beta(1)

  equal to q, less than q, or greater than q, in straight line
  regression. See Box 3.7.4 in the textbook.

*******************************************************************

  Input the column number that contains the Y data.
DATA > 1

  Input the column number that contains the X data.
DATA > 2

  Input the values of a(0) and a(1) in this order.
DATA > 1 0

  Input the value of q.
DATA > 0

===================================================================

  The point estimate of theta is       0.104583

  Its standard error is      0.0605082

  The computed test statistic t_C is        1.72841

  If the test is NH: theta = q versus AH: theta not = q

     then the P-value is      0.0967580
```

```
If the test is NH: theta < or = q versus AH: theta > q

   then the P-value is      0.0483790

If the test is NH: theta > or = q versus AH: theta < q

   then the P-value is       0.951621
-------------------------------------------------------------------
```

Since the test that we are interested in is

$$\text{NH: } \beta_0 = 0 \text{ versus AH: } \beta_0 \neq 0$$

the P-value is 0.097. If you use $\alpha = 0.05$ for this test then NH is not rejected.

Problems

M3.7.1 This is the problem discussed in Task 3.7.2 where an investigator is interested in testing

$$\begin{aligned} &\text{NH: } 6\beta_0 + 264\beta_1 \leq 50 \ \text{ against} \\ &\text{AH: } 6\beta_0 + 264\beta_1 > 50 \end{aligned}$$

Use the macro **test** to perform this test and determine the P-value. The data are stored in the file **crystal.mtp** on the data disk.

M3.7.2 This problem refers to Problem 3.7.1 in the textbook. The data are stored in the file **shelflif.mtp** on the data disk. Use the macro **test** to determine the P-values for the following tests.

(a) NH: $\beta_1 = 0$ versus AH: $\beta_1 \neq 0$.

(b) NH: $\mu_Y(13) \geq 650$ versus AH: $\mu_Y(13) < 650$.

3.8 Analysis of Variance

In Section 3.8 of the textbook we explain **analysis of variance** and discuss the analysis of variance (ANOVA) table. The Minitab command `regress` will automatically produce an analysis of variance table unless `brief` is set to 0. Recall that the default value for `brief` is 2.

To illustrate, we use Example 3.8.1. The age and blood pressure data in this example are given in Table 3.6.2 and are also stored in the file **agebp.mtp**. You should first retrieve this file and examine its contents. We regress Y (blood pressure) which is in column `c1`, on age (X), which is in column c2. The command is

```
regress c1 1 c2
```

Minitab responds with:

```
--------------------------------------------------------------------------

The regression equation is
bp = 66.8 + 1.61 age

Predictor        Coef        Stdev    t-ratio        p
Constant       66.808        2.200      30.37    0.000
age           1.60853      0.04720      34.08    0.000

s = 2.836        R-sq = 98.1%     R-sq(adj) = 98.1%

Analysis of Variance

SOURCE         DF          SS          MS          F          p
Regression      1      9337.7      9337.7    1161.31      0.000
Error          22       176.9         8.0
Total          23      9514.6
--------------------------------------------------------------------------
```

Compare this ANOVA table with the one computed and displayed in Table 3.8.2. Note that the P-value given in the ANOVA table above is for the test of NH: $\beta_0 = \beta_1 = 0$ versus AH: at least one of β_0 and β_1 is nonzero.

Problems

M3.8.1 For the shelf-life data in Table 3.7.1, use Minitab commands to produce an analysis of variance table. The data are stored in the file **shelflif.mtp** on the data disk.

M3.8.2 For the power plant SO_2 data in Table 3.5.4, compute and display an ANOVA table. The data are stored in the file **so2.mtp** on the data disk.

M3.8.3 For the grades26 data in Table 3.2.2, compute and display an ANOVA table. These data are stored in the file **grades26.mtp** on the data disk.

3.9 Coefficient of Determination and Coefficient of Correlation

The output of the Minitab command **regr** contains a quantity denoted by `R-sq`, which is a valid point estimate of $\rho^2_{Y,X}$ whenever the sample data are obtained by simple random sampling. If this is so, the square root of `R-sq`, using the sign of $\hat{\beta}_1$, results in an estimate of $\rho_{Y,X}$, the simple correlation coefficient. There is no Minitab command for computing confidence intervals for $\rho_{Y,X}$. However, you can easily obtain 95% or 99% two-sided confidence intervals for $\rho_{Y,X}$ using the point estimate of $\rho_{Y,X}$ and the charts in Table T-7 in Appendix T. The procedure is explained in Box 3.9.2 in the textbook. From a confidence interval for $\rho_{Y,X}$, you can obtain a confidence interval for σ_Y/σ as explained in Box 3.9.3.

3.10 Regression Analysis When There Are Measurement Errors

All computations required in this section have been discussed in previous sections.

3.11 Regression Through the Origin

To perform the calculations for a linear regression model that passes through the origin (i.e., when β_0 is known to be zero and is not included in the model), you can use the Minitab command `regress` with the subcommand `noconstant`. The command and subcommand are (assuming that the Y data are in c1 and the X data are in c2)

COMMAND FOR REGRESSION THROUGH THE ORIGIN

```
MTB > regress c1 1 c2;
SUBC> noconstant.
```

These commands will perform calculations for the regression model

$$\mu_Y(x) = \beta_1 x$$

All other Minitab commands and subcommands proceed as discussed in previous sections.

Problems

M3.11.1 For the gravity data in Table 3.11.1, use the appropriate Minitab commands to compute $\hat{\beta}_1$ and $\hat{\sigma}$ if the regression model is $\mu_Y(x) = \beta_1 x$. These data are stored in the file **gravity.mtp** on the data disk.

M3.11.2 For the gravity data in Table 3.11.1, obtain a 90% confidence interval for β_1.

Chapter 4

Multiple Linear Regression

4.1 Overview

No computing instructions are needed in this section.

4.2 Notation and Definitions

All computing needed in this section has been discussed in preceding chapters.

4.3 Assumptions for Multiple Linear Regression

No computations are required for this section.

4.4 Point Estimation

In this section we describe two methods for obtaining the following:

(a) Point estimates of β_i for $i = 0, 1, \ldots, k$.

(b) Point estimates of $\mu_Y(x_1, \ldots, x_k)$ and $Y(x_1, \cdots, x_k)$.

(c) Point estimate of $\sigma_{Y|X_1,\cdots,X_k}$.

The two methods are

(1) By using the formulas in (4.4.8), (4.4.9), (4.4.10), and (4.4.16) and carrying out the calculations using matrix commands in Minitab.

(2) By using the `regress` command in Minitab.

We use Example 4.4.2 to illustrate these two methods. In that example the value of k is 4, but you should have no trouble doing the computations for any value of k.

Regression Computations Using Matrices

Consider Example 4.4.2 where the data are given in Table 4.4.3 and are also stored in the file **gpa.mtp** on the data disk. The population regression function, given in (4.4.24), is

$$\mu_Y(x_1, x_2, x_3, x_4) = \beta_0 + \beta_1 x_1 + \beta_2 x_2 + \beta_3 x_3 + \beta_4 x_4$$

where Y is GPA at the end of one year, X_1 is SATmath, X_2 is SATverbal, X_3 is HSmath, and X_4 is HSenglish. We explain the Minitab commands for using matrices to obtain the quantities in (a), (b), and (c) above. As usual, you should retrieve the file **gpa.mtp** and examine its contents.

We begin by creating a column, say c6, containing twenty 1's, which will be needed to form the first column of the X matrix. This can be done using the `set` command as follows.

SET COMMAND

```
MTB > set c6
DATA> 20(1)
DATA> end
```

The above command asks Minitab to enter twenty 1's in column c6. You can check this by printing column c6.

Next we copy the appropriate columns of the data in Table 4.4.3 to form the X matrix using the following command.

COMMAND FOR COPYING COLUMNS INTO A MATRIX

```
MTB > copy c6 c2 c3 c4 c5 m1
```

This command forms a matrix, labeled `m1`. The first column of this matrix is column c6; the remaining columns of the matrix are columns c2, c3, c4, and c5, in that order. We print the matrix `m1` using the command `print m1` so that you can view and check it.

Minitab responds with:

```
------------------------------------------------------------------------

MATRIX M1

      1.00   321.00   247.00    2.30    2.63
      1.00   718.00   436.00    3.80    3.57
      1.00   358.00   578.00    2.98    2.57
      1.00   403.00   447.00    3.58    2.21
      1.00   640.00   563.00    3.38    3.48
      1.00   237.00   342.00    1.48    2.14
      1.00   270.00   472.00    1.67    2.64
      1.00   418.00   356.00    3.73    2.52
      1.00   443.00   327.00    3.09    3.20
      1.00   359.00   385.00    1.54    3.46
      1.00   669.00   664.00    3.21    3.37
      1.00   409.00   518.00    2.77    2.60
      1.00   582.00   364.00    1.47    2.90
      1.00   750.00   632.00    3.14    3.49
      1.00   451.00   435.00    1.54    3.20
      1.00   645.00   704.00    3.50    3.74
      1.00   791.00   341.00    3.20    2.93
      1.00   521.00   483.00    3.59    3.32
      1.00   594.00   665.00    3.42    2.70
      1.00   653.00   606.00    3.69    3.52

------------------------------------------------------------------------
```

Next we form the vector y, which is actually a 20×1 matrix, by copying column c1 into a matrix labeled m2. The command is `copy c1 m2` . Then we print the matrix m2 using the command `print m2` .

Minitab responds with:

```
--------------------------------------------------------------------------------

 MATRIX M2

  1.97
  2.74
  2.19
  2.60
  2.98
  1.65
  1.89
  2.38
  2.66
  1.96
  3.14
  1.96
  2.20
  3.90
  2.02
  3.61
  3.07
  2.63
  3.11
  3.20
--------------------------------------------------------------------------------
```

The following set of Minitab commands is used to compute $X^T X$, store the result in the matrix m4, and print it.

```
MTB > transpose m1 m3
MTB > multiply m3 m1 m4
MTB > print m4
```

The `transpose` command transposes the matrix `m1` and puts the result in the matrix `m3`. The `multiply` command multiplies the matrix `m3` with the matrix `m1`, in this order, and puts the result in the matrix `m4`. The matrix `m4` is then printed.

Minitab responds with:

```
-----------------------------------------------------------------

 MATRIX M4

      20.0     10232.0      9565.0        57.1        60.2
   10232.0   5759520.0   5084391.0     30707.7     31806.9
    9565.0   5084391.0   4908617.0     28231.5     29294.3
      57.1     30707.7     28231.5       176.7       173.9
      60.2     31806.9     29294.3       173.9       185.6
-----------------------------------------------------------------
```

We then compute the matrix $X^T y$, store the result in the matrix `m5`, and print it. The commands are

```
MTB > multiply m3 m2 m5
MTB > print m5
```

Minitab responds with:

```
-----------------------------------------------------------------

 MATRIX M5

      51.9
   28199.6
   25825.6
     155.0
     159.5
-----------------------------------------------------------------
```

Thus `m5` is the matrix $X^T y$.

Note: When doing matrix arithmetic using Minitab, you must remember what the matrices `m1`, `m2`, etc., represent. For example, `m1` is X, `m2` is y, etc.

The following commands are used to invert the matrix m4 (which is actually $X^T X$) and store the result in m6.

INVERT COMMAND

```
MTB > invert m4 m6
```

The matrix m6 is then printed using the command `print m6` .

Minitab responds with:

```
--------------------------------------------------------------------------

 MATRIX M6

  2.65513    0.00138  -0.00036  -0.20069  -0.85213
  0.00138    0.00000  -0.00000  -0.00036  -0.00086
 -0.00036   -0.00000   0.00000  -0.00020  -0.00030
 -0.20068   -0.00036  -0.00020   0.11706   0.04722
 -0.85213   -0.00086  -0.00030   0.04722   0.43205
--------------------------------------------------------------------------
```

To print the matrix m6, Minitab has rounded each element in the matrix to five decimal places. This may not be accurate enough, and it may be advisable to have a more accurate version of this matrix printed out. To do this, you need to copy the five columns of the matrix into some five columns of the worksheet, say columns c11–c15, and you must print the elements of these columns using the `print` command and the `format` subcommand. The following commands show you how to do this.

FORMAT SUBCOMMAND FOR PRINT COMMAND

```
MTB > copy m6 c11-c15
MTB > print c11-c15;
SUBC> format(5f15.10).
```

The `copy` command copies the columns of the matrix m6 into the columns c11–c15, in order. The `print` command asks Minitab to print these columns according to the specifications given in the `format` subcommand. In that subcommand, the instruction `5f15.10` is interpreted as follows. The number 5 in this instruction tells Minitab that there are 5 columns to be printed. The letter `f` stands for *floating point numbers.* The number 15 before the decimal point . asks Minitab to use 15 spaces for each number to be printed with *leading blanks* if necessary. The number 10 after the decimal point tells Minitab to print 10 digits after the decimal point for each number printed. The `format` subcommand is based on the FORTRAN programming language. You should consult the *Minitab Reference Manual* for details. Note that the `format` subcommand is not available with the `print` command when using it to print matrices or constants. It can only be used when printing columns. The Minitab response to the above set of commands is

```
--------------------------------------------------------------------------------
   2.6551275253   0.0013784376  -0.0003619414  -0.2006850839  -0.8521291614
   0.0013784384   0.0000047375  -0.0000004015  -0.0003559387  -0.0008620193
  -0.0003619394  -0.0000004015   0.0000042188  -0.0001953019  -0.0002966869
  -0.2006849051  -0.0003559386  -0.0001953017   0.1170562878   0.0472196266
  -0.8521297574  -0.0008620194  -0.0002966861   0.0472197272   0.4320531487
--------------------------------------------------------------------------------
```

Caution: Even though 10 digits are displayed after the decimal point for each element of the matrix above, the trailing digits may not all be accurate.

Now we compute $\hat{\beta}$ using the formula in (4.4.8), i.e., by computing `m6` times `m5` and putting the result in `m7`. Note that this actually multiplies $(\mathbf{X}^T\mathbf{X})^{-1}$ by $\mathbf{X}^T\mathrm{y}$ on the right. The command is

```
multiply m6 m5 m7
```

The result is then printed using the command `print m7` .

Minitab responds with:

```
--------------------------------------------------------------------------------
 MATRIX M7

 0.161535
 0.002010
 0.001252
 0.189439
 0.087567
--------------------------------------------------------------------------------
```

Note that the elements of m7 are the elements in $\hat{\boldsymbol{\beta}}$, and they are the same as in Exhibit 4.4.1 (except perhaps for rounding errors). The values in Exhibit 4.4.1 were obtained from a Minitab output.

Next we obtain $\hat{e}$, the vector of residuals, by computing the quantity $\boldsymbol{y} - \boldsymbol{X}\hat{\boldsymbol{\beta}}$. The vector $\boldsymbol{y}$ is in m2. The vector $\boldsymbol{X}\hat{\boldsymbol{\beta}}$ is computed and stored in m8. Then m8 is subtracted from m2 to obtain the residual vector $\hat{e}$ (stored in m9). Thus m9 is the matrix (actually a 20×1 vector) $\hat{e}$ in (4.4.18). The commands are

```
multiply m1 m7 m8
subtract m8 m2 m9
```

Then we print the result using the command `print m9` .

Minitab responds with:

```
--------------------------------------------------------------------

 MATRIX M9

 0.187897
-0.443276
-0.204519
 0.196921
-0.118066
 0.116042
 0.047144
 0.005156
 0.332911
 0.000002
-0.100995
-0.424751
-0.119672
 0.539004
-0.164776
 0.279824
 0.028646
-0.154452
 0.037403
-0.040271

--------------------------------------------------------------------
```

Finally we compute $\hat{\sigma}_{Y|X_1,X_2,X_3,X_4} = \hat{\sigma}$. We do this by first calculating the quantity $SSE = \hat{e}^T\hat{e}$ and then dividing it by 15, which is the number of degrees of freedom associated with SSE. To obtain $\hat{e}^T\hat{e}$, we first transpose $\hat{e}$ in m9 and store the result in m10. Then we multiply m10 by m9 (on the right) and store the resulting scalar in k1. The commands are

```
transpose m9 m10
multiply m10 m9 k1
```

Minitab responds with:

```
-----------------------------------------------------------------
ANSWER =           1.0815
-----------------------------------------------------------------
```

Then we divide k1 by 15 and put the result in k2. Thus k2 contains the value of $\hat{\sigma}^2$. Finally we compute the square root of k2 and store it in k3. Thus k3 contains the value of $\hat{\sigma}$. The commands are

```
let k2=k1/15
let k3=sqrt(k2)
```

The result is then printed using the command `print k3` .

Minitab responds with:

```
-----------------------------------------------------------------
K3       0.268514
-----------------------------------------------------------------
```

So we have $\hat{\sigma} = 0.268514$.

Although the above calculations were carried out to illustrate the use of the matrix formulas presented in Chapter 4, in practice there is no need to carry out these calculations since Minitab has certain built-in commands that automatically perform the necessary matrix computations for multiple linear regression. We discuss these next.

The REGRESS Command

The primary Minitab command for computations in multiple regression is the `regress` command, which is similar to the `regress` command used in Chapter 3.

For an illustration, suppose we want to obtain the estimated regression function of Y on X_1, X_2, X_3, and X_4. Further suppose that the Y values are in column c1, and the values of X_1, X_2, X_3, and X_4 are in columns c2, c3, c4, and c5, respectively. Then the `regress` command is

COMMAND FOR REGRESSION

```
MTB > regress c1 4 c2 c3 c4 c5
```

The first column after the command `regress`, namely column c1, is the column that contains the response variable Y. The next quantity, namely the number 4, is the number of predictor variables. The final columns, namely, c2, c3, c4, and c5, are the columns that contain the four predictor variables.

Minitab responds with:

```
---------------------------------------------------------------------

The regression equation is
GPA = 0.162 + 0.00201 SATmath + 0.00125 SATverb +
      0.189 HSmath + 0.088 HSengl


Predictor        Coef        Stdev     t-ratio       p
Constant       0.1615       0.4375        0.37   0.717
SATmath     0.0020102    0.0005844        3.44   0.004
SATverb     0.0012522    0.0005515        2.27   0.038
HSmath        0.18944      0.09187        2.06   0.057
HSengl         0.0876       0.1765        0.50   0.627

s = 0.2685      R-sq = 85.3%      R-sq(adj) = 81.4%
---------------------------------------------------------------------
```

The actual output following the `regress` command contains additional information that is not needed at this point, so we have given here only that portion of the output that is related to point estimation. This output is the same as that in Exhibit 4.4.1.

Thus $\hat{\beta}_0 = 0.1615$, $\hat{\beta}_1 = 0.0020102$, $\hat{\beta}_2 = 0.0012522$, $\hat{\beta}_3 = 0.18944$, $\hat{\beta}_4 = 0.0876$, and $\hat{\sigma} = 0.2685$. The estimated regression function of Y on X_1, X_2, X_3, X_4 is

$$\hat{\mu}_Y(x_1, x_2, x_3, x_4) = 0.1615 + 0.0020102x_1 + 0.0012522x_2 + 0.18944x_3 + 0.0876x_4$$

Problems

Problems **M4.4.1–M4.4.8** refer to Task 4.4.1. The data are given in Table 4.4.4 and are also stored in the file **table444.mtp** on the data disk.

M4.4.1 Retrieve these data and print them.

M4.4.2 Give the Minitab commands to obtain and exhibit each of the following matrices: $\boldsymbol{X}$, $\boldsymbol{y}$, $\boldsymbol{X}^T\boldsymbol{X}$, $(\boldsymbol{X}^T\boldsymbol{X})^{-1}$, $\boldsymbol{X}^T\boldsymbol{y}$.

M4.4.3 Give the Minitab commands to compute $\hat{\boldsymbol{\beta}}$ and exhibit it.

M4.4.4 Give the Minitab commands to compute $\hat{\boldsymbol{e}} = \boldsymbol{y} - \boldsymbol{X}\hat{\boldsymbol{\beta}}$ and exhibit its value.

M4.4.5 Give the Minitab commands to compute $\hat{\sigma}_{Y|X_1,X_2}$ and exhibit its value.

M4.4.6 Use the `regress` command in Minitab to obtain $\hat{\boldsymbol{\beta}}$ and $\hat{\sigma}$.

M4.4.7 Suppose the regression function of Y on X_1 is given by $\mu_Y^{(A)}(x_1) = \beta_0^A + \beta_1^A x_1$. Find $\hat{\beta}_0^A$, $\hat{\beta}_1^A$, and $\hat{\sigma}_{Y|X_1}$ using matrix commands.

M4.4.8 In Problem **M4.4.7**, compute the quantities using the `regress` command.

4.5 Residual Analysis

In this section we discuss how Minitab can be used to perform the calculations needed for residual analysis of Section 4.5. Specifically, we discuss Minitab commands that can be used to compute residuals, fits, standardized residuals, and nscores for multiple linear regression. We use the electric bill data in Example 4.5.1 as an illustration. The data are displayed in Table 4.5.1 and are also stored in the file **electric.mtp** on the data disk. As usual, you should first retrieve the data from the file **electric.mtp** and check them. We note that column `c1` contains the response variable, `Bill`, and columns `c2`, `c3`, and `c4` contain the predictor variables, `Income`, `Persons`, and `Area`, respectively.

An extended version of the Minitab command `regress` can be used to obtain the standardized residuals r_i and the fitted values in addition to the point estimates of β_0, β_1, β_2, and β_3. Recall that the basic command for obtaining the sample regression function of `Bill` (which is in `c1`) on `Income`, `Persons`, and `Area` (which are in `c2`, `c3`, and `c4`, respectively) is `regress c1 3 c2 c3 c4` , where the first argument is the column containing the values of the response variable, viz., c1 for this data set; the second argument is k, the number of predictor variables in the regression function, which for this problem is k=3; and the third set of arguments is the list of columns containing the values for the predictor variables, which for this problem is c2 c3 c4.

In the extended version of the `regress` command you may specify two additional columns, which are used by Minitab to store the standardized residuals and the fitted values respectively. Suppose the two additional columns we specify as part of the `regress` command are columns c5 and c6. Then the standardized residuals will be stored in the first of the two optional columns, viz., c5, and the fitted values will be stored in the second of the optional columns, viz., c6. If you also want to compute the residuals $\hat{e}_i$, you can use the subcommand `residuals c7` , which puts the residuals in column c7. You will note that these commands are very similar to the commands for straight line regression discussed in Section 3.5 of this manual. The command and the subcommand are

EXTENDED REGRESS COMMAND AND RESIDUALS SUBCOMMAND

```
MTB > regress c1 3 c2 c3 c4 c5 c6;
SUBC> residuals c7.
```

Minitab responds with:

```
-----------------------------------------------------------------------------

The regression equation is
Bill = - 358 + 0.075 Income + 55.1 Persons + 0.281 Area

Predictor        Coef        Stdev    t-ratio        p
Constant       -358.4        198.7      -1.80    0.081
Income         0.0751       0.1361       0.55    0.585
Persons         55.09        29.05       1.90    0.068
Area           0.2811       0.2261       1.24    0.223

s = 135.4       R-sq = 85.1%     R-sq(adj) = 83.7%
```

```
Analysis of Variance

SOURCE        DF        SS          MS         F        p
Regression     3   3151505     1050502     57.28    0.000
Error         30    550163       18339
Total         33   3701668
------------------------------------------------------------------
```

This output is part of Exhibit 4.5.1. Note that the standardized residuals, the fits, and the residuals are not printed automatically. They may be printed using the command `print c5 c6 c7`.

Next we obtain the nscores of the standardized residuals (which are in column c5) and put them in column c8. The Minitab command is

NSCORE COMMAND

```
MTB > nscores c5 c8
```

The next set of commands instructs Minitab to name columns c5-c8 as follows.

COMMAND TO NAME COLUMNS

```
MTB > name c5 'stdresid'
MTB > name c6 'fits'
MTB > name c7 'residual'
MTB > name c8 'nscores'
```

You should print the data values, the fitted values, the residuals, the standardized residuals, and the nscores, which are in columns c1 to c8, respectively, and compare them with the corresponding results in Exhibit 4.5.1. Also, using the above results you can obtain the plots in Example 4.5.1.

Checking Gaussian Assumptions

Next we exhibit Minitab commands that can be used to help decide whether or not the four-variable population $\{(Y, X_1, X_2, X_3)\}$ is Gaussian where Y = `Bill`, X_1 = `Income`, X_2 = `Persons`, and X_3 = `Area`. For this purpose we need to examine several linear combinations of these four variables, including the variables Y, X_1, X_2, and X_3 themselves. We consider four different linear combinations of the variables Y, X_1, X_2, and X_3, which are in columns c1, c2, c3, and c4, respectively, in the file **electric.mtp** on the data disk. Retrieve this file first. The linear combinations are computed and the results stored in columns c11, c12, c13, and c14, respectively, using the following Minitab commands.

COMMANDS TO OBTAIN LINEAR COMBINATIONS OF VARIABLES

```
MTB > let c11= 5*c1+c2+1500*c3+2*c4
MTB > let c12= 5*c1+c2-1500*c3+2*c4
MTB > let c13= 5*c1+c2-1500*c3-2*c4
MTB > let c14= -5*c1+c2-1500*c3-2*c4
```

The coefficients which make up the linear combinations being examined are chosen so that no single variable dominates the value of the linear combination. This is especially important when the different variables forming the linear combinations take on values that are not commensurate with each other as is the case here – the sample values of Y (Bill) range between $96.00 and $1,272.00 whereas the sample values for the variable X_2 (Persons) range between 1 and 7. You should try several other linear combinations.

We now compute the nscores for each of the four linear combinations considered, and construct the corresponding rankit-plots. The commands to compute the nscores are as follows.

COMMANDS TO COMPUTE NSCORES

```
MTB > nscore c11 c21
MTB > nscore c12 c22
MTB > nscore c13 c23
MTB > nscore c14 c24
```

The nscores for c11, c12, c13, and c14 are stored in c21, c22, c23, and c24, respectively. We now obtain the various rankit-plots. For a rankit-plot of c11, the values in c11 are plotted against the corresponding nscores in c21 using the command plot c11 c21 .

Minitab responds with:

```
----------------------------------------------------------------------------


        -                                                       *
C11     -
        -                                              *  *
        -                                            *
   21000+                                         * *
        -                                     * **
        -                                    *
        -                                 ***
        -                              ***
   14000+                          *2 2
        -                         *
        -                      ***
        -                 *** *
        -             * *
    7000+   *    *  *
        -
        -

         --------+---------+---------+---------+---------+--------C21
              -1.60     -0.80      0.00      0.80      1.60
----------------------------------------------------------------------------
```

Similarly, by using the commands

```
plot c12 c22
plot c13 c23
plot c14 c24
```

we obtain the remaining 3 rankit-plots. They are given below in order.

```
-------------------------------------------------------------------------------
         -
    15000+
         -                                                       *
 C12     -                                         * * * *  *
         -
         -                                        *
    10000+                                       *
         -                                   ***
         -                                 **
         -                             2***
         -                         ****
     5000+                     ****
         -             * * ***
         -
         -
         -           *
        0+   *    *
          --------+---------+---------+---------+---------+--------C22
              -1.60     -0.80      0.00      0.80      1.60

     3500+                                                       *
         -                                                  *
 C13     -                                               *
         -                                    ** *** * *
         -                                ****
        0+                            *2**
         -                        ****
         -                      **
         -                  ** *
         -                 *
    -3500+           * * *
         -        *
         -
         -
         -
    -7000+   *
         -
          --------+---------+---------+---------+---------+--------C23
              -1.60     -0.80      0.00      0.80      1.60
```

```
        -                                                       *
C14     -                                             * *  *
        -                                        ** *
        -                                  *** 2
   -5000+                                **
        -                           **2**
        -                        ***
        -                     ***
        -                   *
  -10000+                 **
        -               *
        -
        -
        -             *
  -15000+   *    *  *
        -
        -

          --------+---------+---------+---------+---------+--------C24
              -1.60     -0.80      0.00      0.80      1.60
```

The last of these rankit-plots seems to suggest that this particular linear combination of the four variables Y, X_1, X_2, and X_3 may not form a Gaussian population because the graph of this linear combination against the rankit values appears to exhibit some curvature. So the four-variable population $\{(Y, X_1, X_2, X_3)\}$ may not be Gaussian. Hence the validity of assumptions **(B)** in this example are questionable.

Problems

M4.5.1 For Problem 4.5.2 in the textbook, use Minitab commands and subcommands discussed in this section to work **(a)** through **(f)**.

(a) Regress Y on X_1, X_2, X_3, and X_4 where Y = GPA, X_1 = SATmath, X_2 = SATverb, X_3 = HSmath, and X_4 = HSengl.

(b) Compute the fits $\hat{\mu}_Y(x_1, x_2, x_3, x_4)$.

(c) Compute the residuals $\hat{e}_i$.

(d) Compute the standardized residuals r_i.

(e) Compute the nscores $z_i^{(n)}$ of the standardized residuals.

(f) Plot the standardized residuals against the fits, against the Y values, against the nscores, and against each of X_1, X_2, X_3, and X_4.

M4.5.2 In Problem **M4.5.1**, obtain rankit-plots of several linear combinations of the variables Y, X_1, X_2, X_3, and X_4.

M4.5.3 From the results of Problem **M4.5.2**, do you have any reason to believe that the five-variable population $\{(Y, X_1, X_2, X_3, X_4)\}$ is not Gaussian?

4.6 Confidence Intervals

Sections 4.4 and 4.6 give formulas for computing $\hat{\beta}_i$ for $i = 0, 1, 2, \ldots, k$ and standard errors of the $\hat{\beta}_i$, which are the ingredients needed to compute confidence intervals for the β_i. In Section 4.4 of this manual we showed how these quantities can be obtained using the Minitab command `regress` . However, the only confidence intervals that Minitab will compute directly are two-sided 95% confidence intervals for $\mu_Y(x_1, \ldots, x_k)$ and $Y(x_1, \cdots, x_k)$. If you select $x_1 = x_2 = \ldots = x_k = 0$ when computing confidence limits for $\mu_Y(x_1, \ldots, x_k)$, then you can obtain confidence intervals for β_0 since $\mu_Y(0, 0, \ldots, 0) = \beta_0$. There is no command in Minitab that can be used to compute estimates of general linear combinations of the β_i, or the standard errors of these estimates. In Section 4.6 you learned how these can be computed by using matrices, but this requires a significant amount of tedious work. To make it easier to obtain point estimates, their standard errors, and confidence intervals for $\sum_{i=0}^{k} a_i\beta_i$, for values of a_i that you specify, we have written a macro that computes these quantities. In this section we show how to use this macro. The Minitab commands for the macro are stored in the file **cilinear.mtb**, which stands for confidence interval for **linear** combinations of the β_i. Two subprograms stored in the files **cilinr0.mtb** and **cilinr1.mtb** are needed to run this macro. All three files are in the subdirectory macro on the data disk.

We illustrate this macro by computing a 90% two-sided confidence interval for β_3 using the GPA data as required in part 1 of Task 4.6.1. As usual, first retrieve the file **gpa.mtp** and examine its contents. The command to start the macro is

COMMAND TO START THE MACRO CILINEAR

```
MTB > execute 'b:\macro\cilinear'
```

Minitab responds with:

```
-------------------------------------------------------------------------

  This macro will compute a confidence interval for theta where
  theta is any linear combination
         a(0) beta(0) + a(1) beta(1) + ... + a(k) beta(k)
  of the parameters beta(0), ..., beta(k).

  Confidence intervals for beta(0), beta(1), ..., beta(k), and
  mu_Y(x(1), ..., x(k)) may be obtained as special cases by properly
  choosing a(0), a(1), ..., a(k).

  You are required to input the coefficients of the vector  a,
  which has k+1 components, viz., a(0), a(1), ..., a(k). For a
  model without an intercept, only the k constants a(1), ..., a(k)
  are required.

  You are also required to specify the confidence coefficient.

  Leave columns c80-c100 empty for work space.

  The data must be organized such that X(1), X(2), ..., X(k) are
  in consecutive columns in the order indicated.

***********************************************************************

  Input the column number for the column containing Y.
DATA> 1

  Input the column number for the column containing X(1).
DATA> 2

  Input the value of k, the number of predictors in your model.
DATA> 4

  Does your model have an intercept term? If yes, input 1, if no,
  input 0.
DATA> 1

  Input the constants a(0), a(1), a(2), ..., a(k) in the linear
  function of interest in the indicated order. If your model has
  no intercept, then input a(1), ..., a(k) only.
DATA> 0 0 0 1 0
```

```
  Input the confidence coefficient for a two-sided confidence
  interval.
DATA> .90
======================================================================

  The estimated value of theta is    0.1894402

  The standard error of this estimate is   0.09186804

  For a two-sided confidence interval for theta with
                                  confidence coefficient  0.900

  the lower confidence bound is   0.02839059   and

  the upper confidence bound is    0.3504898.
----------------------------------------------------------------------
```

Thus we get $\hat{\beta}_3 = 0.18944$ and the confidence statement is

$$C[0.0284 \leq \beta_3 \leq 0.3505] = 0.90$$

These values are of course the same as obtained in Task 4.6.1.

In some problems, a confidence interval for σ is the quantity of interest. This can be obtained by using the macro **sgmaconf.mtb** discussed in Section 3.6 of this manual. The degrees of freedom associated with $\hat{\sigma}$ are $n - k - 1$ if the model has an intercept term, and $n - k$ if the model does not have an intercept term.

Problems

M4.6.1 Use Minitab commands to work the problems in Section 4.6 of the textbook. Use macros when Minitab commands are not available.

4.7 Tests

In Section 3.7 of this manual we explained how to use a Minitab macro to do the computing needed for statistical tests in straight line regression discussed in Boxes 3.7.1 to 3.7.4 in the textbook. In this section we describe two macros that can be used to perform the tests described in Boxes 4.7.1 and 4.7.2. The first macro, called **testmult**, computes the test statistic t_C and the P-value for tests explained in Box 4.7.1 about linear combinations $a_0\beta_0 + a_1\beta_1 + \cdots + a_k\beta_k$ in multiple regression. The second macro, called **sgmatest**, computes the P-value for tests about the subpopulation standard deviation σ discussed in Box 4.7.2.

The Macro testmult

The Minitab commands for the macro **testmult** are stored in the files **testmult.mtb, testmlt0.mtb** and **testmlt1.mtb**. All three files are stored in the subdirectory macro on the data disk. To use this macro, you need to have the predictor variables, $X_1, X_2, \ldots, X_k$, in consecutive columns, in this order. You will be asked to input the number of the column containing the response variable Y, the number of the column containing the predictor variable X_1, the number of predictors k, the values of $a_0, a_1, \ldots, a_k$, and the value of q. You can also use this macro in straight line regression problems by specifying the value of **k** as 1 (one predictor variable).

We illustrate this macro by using the GPA data in Example 4.7.1. The data are in the file **gpa.mtp** on the data disk. You should first retrieve this file. We test

$$\text{NH}: \ \mu_Y(594, 665, 3.42, 2.70) \leq 2.5 \text{ versus AH}: \ \mu_Y(594, 665, 3.42, 2.70) > 2.5$$

The command to start the macro is

COMMAND FOR COMPUTING TESTS

```
MTB > execute 'b:\macro\testmult'
```

Minitab responds with:

```
This program will compute the test statistic t_C  for testing
a linear combination
    theta = a(0) beta(0) + a(1) beta(1) + ... + a(k) beta(k),
of the beta parameters in multiple linear regression.
```

```
  You are required to input the coefficients of the vector  a
  which has k+1 components,  viz.,  a(0), a(1), ..., a(k).
  For a model without an intercept, only the k constants
  a(1), ..., a(k) are required.

  The data must be organized such that X(1), X(2), ..., X(k) are
  in consecutive columns, in this order.

**********************************************************************

  Input the column number for the column containing Y.
DATA> 1

  Input the column number for the column containing X(1).
DATA> 2

  Input k, the number of predictors in your model.
DATA> 4

  Does your model have an intercept term. If yes, input 1, if no,
  input 0.
DATA> 1

  Input the constants a(0), a(1), a(2), ..., a(k) in the linear
  function of interest, in this order. If your model has no
  intercept, then only input a(1), ..., a(k).
DATA> 1 594 665 3.42 2.70

  Input the value of q.
DATA> 2.5

======================================================================

  The estimated value of theta is     3.072608

  The standard error of this estimate is    0.1428586

  The test statistic t_C is     4.008214

  If the test is NH: theta = q against AH: theta not equal to q
  then the P-value is  0.001142025
```

```
If the test is NH: theta < or = q against AH: theta > q
then the P-value is  0.000571012

If the test is NH: theta > or = q against AH: theta < q
then the P-value is    0.9994290
```

So the P-value for the test of interest is 0.00057 (rounded to 5 decimals). You should check the above results against the calculations shown in Example 4.7.1.

The Macro **sgmatest**

You can use the macro **sgmatest** to carry out tests concerning subpopulation standard deviations discussed in Box 4.7.2. The macro can even be used to carry out tests about σ_Y in a one-variable population $\{Y\}$. The Minitab commands required for this macro are stored in the file **sgmatest.mtb** in the subdirectory `macro` on the data disk.

We illustrate the use of this macro by carrying out a test of

$$\text{NH}: \ \sigma \geq 0.2 \ \text{ against } \ \text{AH}: \ \sigma < 0.2$$

in Example 4.7.1. Refer to part (b) of that example. The command to start the macro **sgmatest** is

COMMAND FOR COMPUTING TESTS ABOUT STANDARD DEVIATIONS

```
MTB > execute 'b:\macro\sgmatest'
```

Minitab responds with:

```
This program will compute the test statistic and the P-value
for a statistical test of

NH: sigma      = q   against   AH: sigma not = q,
NH: sigma < or = q   against   AH: sigma > q,     and
NH: sigma > or = q   against   AH: sigma < q
```

```
  Here sigma may be the standard deviation of a one-variable
  population {Y} or the standard deviation in a regression problem.

  You must input the estimated value of the standard deviation
  being tested. You must also input the degrees of freedom for
  estimating sigma and the value of q.

************************************************************************

  Input the estimate of sigma
DATA> .2685

  Input the degrees of freedom associated with the estimate of sigma
DATA> 15

  Input the value of q
DATA> .2
========================================================================

  The test statistic Chisquare_C is      27.03459

  For a test of NH: sigma = q against AH: sigma not = q

                                        the P-value is   0.05689216

  For a test of NH: sigma < or = q against AH: sigma > q

                                        the P-value is   0.02844608

  For a test of NH: sigma > or = q against AH: sigma < q

                                        the P-value is    0.9715539

------------------------------------------------------------------------
```

Thus the P-value for this problem is 0.972 (rounded to three decimals).

Problems

M4.7.1 For the GPA data in the file **gpa.mtp** on the data disk, test the following using $\alpha = .05$. Give your conclusion for each.

(a) NH : $\beta_1 = 0.003$ versus AH : $\beta_1 \neq 0.003$.

(b) NH : $\beta_2 \leq 0.001$ versus AH : $\beta_2 > 0.001$.

(c) NH : $\mu_Y(500, 615, 3.10, 2.90) \leq 2.5$ versus AH : $\mu_Y(500, 615, 3.10, 2.90) > 2.5$.

M4.7.2 Work part (b) of Problem 4.7.1 in the textbook using one of the macros discussed in this section.

M4.7.3 Work Problem 4.7.3 in the textbook using one of the macros discussed in this section.

4.8 Analysis of Variance

In Section 4.8 of the textbook we discussed the quantities displayed in an analysis of variance table and how these quantities can be used in regression. An ANOVA table can be computed with the Minitab command `regress` , and we demonstrate this by using Example 4.8.1. The data are given in Table 4.4.3 and are stored in the file **gpa.mtp** on the data disk. Retrieve this file and issue the following command.

REGRESSION COMMAND

```
MTB > regress c1 4 c2 c3 c4 c5
```

Minitab responds with:

```
-------------------------------------------------------------------------

The regression equation is
GPA = 0.162 + 0.00201 SATmath + 0.00125 SATverb
            + 0.189 HSmath + 0.088 HSengl

Predictor        Coef        Stdev     t-ratio        p
Constant       0.1615       0.4375        0.37    0.717
SATmath     0.0020102    0.0005844        3.44    0.004
SATverb     0.0012522    0.0005515        2.27    0.038
HSmath        0.18944      0.09187        2.06    0.057
HSengl         0.0876       0.1765        0.50    0.627

s = 0.2685        R-sq = 85.3%      R-sq(adj) = 81.4%
```

```
Analysis of Variance

SOURCE        DF          SS          MS         F        p
Regression     4      6.2643      1.5661     21.72    0.000
Error         15      1.0815      0.0721
Total         19      7.3458
-----------------------------------------------------------------
```

The quantities in the ANOVA table above are of course the same as the quantities in Table 4.8.2. You should recall that if the value of `brief` is set equal to 0, then the ANOVA will not be printed by Minitab. The default value for `brief` is 2.

4.9 Comparison of Two Regression Functions (Nested Case) and Coefficients of Determination

Since Minitab does not directly compute confidence intervals for ratios of standard deviations, we have supplied a macro on the data disk for this purpose. Suppose we have two models, model A and model B, given by

$$\mu_Y^{(A)}(x_1, \ldots, x_k) = \beta_0^A + \beta_1^A x_1 + \cdots + \beta_k^A x_k \text{ with standard deviation } \sigma_A,$$

and

$$\mu_Y^{(B)}(x_1, \ldots, x_m) = \beta_0^B + \beta_1^B x_1 + \cdots + \beta_m^B x_m \text{ with standard deviation } \sigma_B.$$

Thus model A is the full model and model B is a submodel (i.e., $m < k$).

The program, which is in the file **ratiosgm.mtb** (i.e., **ratio of sigmas**), will compute the following.

- A confidence interval for σ_A, the subpopulation standard deviation for model A, with confidence coefficient $1 - \alpha(A)$ specified by the investigator.
- A confidence interval for σ_B, the subpopulation standard deviation for model B, with confidence coefficient $1 - \alpha(B)$ specified by the investigator.
- A confidence interval for σ_B/σ_A with confidence coefficient greater than or equal to $1 - \alpha(A) - \alpha(B)$.

You must input the following:

(1) The sum of squared errors for model A, i.e., $SSE(A)$, the degrees of freedom associated with $SSE(A)$, and the confidence coefficient $1 - \alpha(A)$ for σ_A.

(2) The sum of squared errors for model B, i.e., $SSE(B)$, the degrees of freedom associated with $SSE(B)$, and the confidence coefficient $1 - \alpha(B)$ for σ_B.

We refer to Example 4.9.4 to illustrate the use of this macro. In that example we want to determine how good model A is for predicting Y (compute a point estimate and a confidence interval for σ_A), how good model B is for predicting Y (compute a point estimate and a confidence interval for σ_B), and how much better model A is than model B for predicting Y (compute a point estimate and a confidence interval for σ_B/σ_A). The two models are

$$\text{model A}: \quad \mu_Y^{(A)}(x_1, x_2, x_3, x_4) = \beta_0^A + \beta_1^A x_1 + \beta_2^A x_2 + \beta_3^A x_3 + \beta_4^A x_4 \quad \text{and}$$
$$\text{model B}: \quad \mu_Y^{(B)}(x_3, x_4) = \beta_0^B + \beta_3^B x_3 + \beta_4^B x_4$$

where Y = GPA, X_1 = SATmath, X_2 = SATverb, X_3 = HSmath, and X_4 = HSengl. The following quantities are given in Exhibit 4.9.1.

(1) $SSE(A)$=1.0815 with 15 degrees of freedom,

(2) $SSE(B)$=2.4180 with 17 degrees of freedom.

As you know, these quantities can be computed using appropriate `regress` commands.

Suppose we want 90% confidence intervals for σ_A and σ_B, i.e., $\alpha(A) = 0.10$ and $\alpha(B) = 0.10$. This will lead to a confidence interval for σ_B/σ_A with confidence coefficient greater than or equal to $1 - \alpha(A) - \alpha(B) = 0.80$. You should first retrieve the file **gpa.mtp**. The Minitab command to start the macro is

COMMAND FOR COMPUTING CONFIDENCE INTERVALS FOR SIGMA(A), SIGMA(B), AND THE RATIO

```
MTB > execute 'b:\macro\ratiosgm'
```

Minitab responds with:

```
------------------------------------------------------------------------

   This program will compute a two-sided confidence interval for
   the following

   (1) sigma(A), the subpopulation standard deviation for model A,
   (2) sigma(B), the subpopulation standard deviation for model B,
   (3) sigma(B)/sigma(A),  the ratio of the standard deviation of
       model B to the standard deviation of model A.

   You must input the quantities SSE(A) and SSE(B), the sum of
   squared errors for model A and model B respectively, and the
   degrees of freedom for estimating sigma(A) and for estimating
   sigma(B).

   You must also input the desired confidence coefficient for
   two-sided confidence intervals for sigma(A) and sigma(B). If the
   confidence coefficient for sigma(A) is 1-alpha(A) and the
   confidence coefficient for sigma(B) is 1-alpha(B), then the
   confidence coefficient for sigma(B)/sigma(A) is greater than or
   equal to 1-alpha(A)-alpha(B).

   The Bonferroni method is used to obtain the confidence interval
   for sigma(B)/sigma(A).

 **********************************************************************

   Input the quantity SSE(A), the sum of squared errors for model A.
 DATA> 1.0815

   Input the degrees of freedom associated with the sum of squared
   errors for model A.
 DATA> 15

   Input the confidence coefficient for sigma(A).
 DATA> .90

   Input the quantity SSE(B), the sum of squared errors for model B.
 DATA> 2.4180

   Input the degrees of freedom associated with the sum of squared
   errors for model B.
 DATA> 17
```

```
   Input the confidence coefficient for sigma(B).
DATA> .90

======================================================================

   The estimate of sigma(A) is    0.2685145

   For a two-sided confidence interval for sigma(A)
                        with confidence coefficient  0.900
   the lower bound is    0.2080079   and
   the upper bound is    0.3859373.

   The estimate of sigma(B) is    0.3771409

   For a two-sided confidence interval for sigma(B)
                        with confidence coefficient  0.900
   the lower bound is    0.2960570   and
   the upper bound is    0.5280493.

   The estimate of sigma(B)/sigma(A) is     1.404546

   For a two-sided confidence interval for sigma(B)/sigma(A)
   with confidence coefficient greater than or equal to  0.800
   the lower bound is    0.7671117   and
   the upper bound is     2.538602
----------------------------------------------------------------------
```

So we get

$$\hat{\sigma}_A = 0.268515, \quad C[0.2080 \leq \sigma_A \leq 0.3859] = 0.90$$

$$\hat{\sigma}_B = 0.377141, \quad C[0.2961 \leq \sigma_B \leq 0.5280] = 0.90$$

$$\frac{\hat{\sigma}_B}{\hat{\sigma}_A} = 1.404546, \text{ and } C[0.7671 \leq \sigma_B/\sigma_A \leq 2.5386] \geq .80$$

Problems

4.9.1 Work Problem 4.9.1 (a) and (b) in the textbook using the macro discussed in this section.

4.9.2 In Problem 4.9.1 in the textbook, compute a two-sided confidence interval for σ_B/σ_A with confidence coefficient greater than or equal to 95%.

4.10 Comparison of Two Multiple Regression Models (Non-nested Case)

To compute confidence intervals for σ_A, σ_B, and σ_B/σ_A for the non-nested case, you can use the macro **ratiosgm** discussed in Section 4.9 of this manual.

4.11 Lack-of-Fit

As explained in Section 4.11, a computer is a practical necessity for computing confidence intervals for the lack-of-fit constants θ_i since tedious matrix calculations are involved. We have written a macro named **lackfit** for this purpose. Besides computing confidence intervals for the lack-of-fit constants, this macro will also output the P-value for a traditional lack-of-fit test. Minitab instructions for the macro are stored in the files **lackfit.mtb**, **lackfit1.mtb**, and **lackfit2.mtb** in the subdirectory macro on the data disk. In this section we explain how to use this macro. We refer to Example 4.11.3 where an investigator wants to examine the relationship between blood pressure (Y) and age (X) to determine whether the model $P_Y(x) = \beta_0 + \beta_1 x$ is close enough to the unknown regression function $\mu_Y(x)$ to be useful for predicting blood pressure based on age. The data are stored in the file **bp.mtp** on the data disk and are exhibited in Table 4.11.2.

To use the macro **lackfit**, you must have the response variable Y in column c1 and the predictor variable X in column c2. We now show how to use the macro for computing point estimates and 95% simultaneous confidence intervals for the θ_i in Example 4.11.3. Retrieve the data from the file **bp.mtp** and start the macro with the following command.

LACK-OF-FIT MACRO

```
MTB > execute 'b:\macro\lackfit'
```

> **Note:** During the execution of the macro, Minitab will periodically pause, and issue the prompt `CONTINUE?` or the prompt `CONTINUE? (y for YES)` so that you will have ample time to read the output on the display. When you are ready to continue with the rest of the output, simply type `y` and press `Enter` .

Minitab responds with:

```
----------------------------------------------------------------------

  This program will compute confidence intervals for the lack-of-fit
  constants theta(1), theta(2), ..., theta(m), such that you have
  at least 1-alpha confidence that all intervals are simultaneously
  correct.

  The program will also carry out the traditional lack-of-fit test
  and display F_C and the P-value.

  Column c1 must contain the Y values.
  Column c2 must contain the X values.

  Leave columns c50 to c99 empty for workspace.

**********************************************************************

  Input 1-alpha, the simultaneous confidence coefficient.
DATA> .95

======================================================================

  The pooled estimate of the subpopulation standard deviation is

                            3.765744

  Point estimates, standard errors, and the lower and the upper
  bounds for two-sided confidence intervals for the lack-of-fit
  constants theta_i are given below. The confidence intervals are
  all simultaneously correct with confidence coefficient greater
  than or equal to

         0.950
```

```
Estimate of      Standard        Lower Confidence      Upper Confidence
  theta_i          Error               Bound                  Bound


 1.573334        1.121324           -1.617215               4.763883
-1.003334        1.440531           -5.102135               3.095467
-2.296669        1.415706           -6.324836               1.731498
 1.309998        1.359498           -2.558238               5.178233
0.4166565        1.087077           -2.676449               3.509762

SS(Pure Error)  =       283.6166
df(Pure Error)  =             20
MS(Pure Error)  =       14.18083.

SS(Lack of fit) =       56.89362
df(Lack of fit) =              3
MS(Lack of fit) =       18.96454.

F_C for the traditional lack-of-fit test =       1.337336
Numerator degrees of freedom    =           3
Denominator degrees of freedom =           20
P-value =      0.2903963
--------------------------------------------------------------------------
```

You should compare these results with those obtained in Examples 4.11.3 and 4.11.4.

Problems

M4.11.1 Rework part (b) of Exercise 4.12.2 using the macros discussed in this section. Check your results against your answers obtained earlier without the use of the macros.

Chapter 5

Diagnostic Procedures

5.1 Overview

There are no calculations in this section that require Minitab.

5.2 Outliers

To examine a set of data for outliers as explained in Section 5.2 of the textbook, it is useful to examine the following:

(1) The fitted values $\hat{\mu}_Y(x)$

(2) The residuals $\hat{e}_i = y_i - \hat{\mu}_Y(x_{i,1}, \ldots, x_{i,k})$

(3) The standardized residuals $r_i = \dfrac{y_i - \hat{Y}(x_{i,1}, \ldots, x_{i,k})}{\hat{\sigma}\sqrt{1 - h_{i,i}}}$

(4) The studentized deleted residuals $T_i = \dfrac{y_i - \hat{Y}_{-i}(x_{i,1}, \ldots, x_{i,k})}{\hat{\sigma}_{(-i)}/\sqrt{1 - h_{i,i}}}$

These quantities can be obtained by using the extended **regress** command and appropriate subcommands. We explain these commands by referring to Example 5.2.1, where an investigator is studying the relationship of Y, the insurance premiums, and X_1, X_2, the age of cars and the price of cars, respectively. The data are given in Table 5.2.1 and are stored in the file **premiums.mtp** on the data disk. First retrieve this file and then issue the following command and subcommands.

REGRESS COMMAND AND DIAGNOSTIC SUBCOMMANDS

```
MTB > regress c1 2 c2 c3 c4 c5;
SUBC> residuals c6;
SUBC> tresiduals c7.
```

The command

```
regress c1 2 c2 c3 c4 c5;
```

with the subcommand

```
residuals c6;
```

was explained in Section 4.5 of this manual. The subcommand

```
tresiduals c7.
```

computes the studentized deleted residuals and puts them in column c7. As usual, only the first four letters of these subcommands are required. So you can use `resi` for `residuals` and `tres` for `tresiduals` . We now name the newly created columns using the name command as follows:

```
MTB > name c4 'stdresid'
MTB > name c5 'fits'
MTB > name c6 'residual'
MTB > name c7 'tresid'
```

You should print these quantities and compare them with the entries in the table in Exhibit 5.2.1.

Sometimes it may be advantageous to examine the columns in a rearranged order. For instance, consider the following command.

```
MTB > print c2 c3 c1 c5 c4 c6 c7
```

The output from this `print` command will have the Y column next to the column of `fits` so you can compare the numbers in these two columns easily.

Problems

M5.2.1 Use the appropriate Minitab commands to obtain the table of residuals, fits, etc., in Exhibit 5.2.2.

5.3 Leverages or Hat Values

Hat values can be computed using the `hi` subcommand of the `regress` command as explained in Section 3.5 of this manual.

5.4 Influential Observations – Cook's Distance and DFFITS

Recall that Cook's distance and/or DFFITS can be helpful in determining which (if any) values in a data set are influential observations. Values of Cook's distance and DFFITS can be computed using subcommands of the Minitab `regress` command. We illustrate these subcommands using the artificial data in Table 5.4.1, which are stored in the file **table541.mtp** on the data disk. First retrieve the data from this file. The Minitab commands for obtaining Cook's distance and DFFITS are

COMMANDS FOR COMPUTING COOK'S DISTANCE AND DFFITS

```
MTB > regress c1 1 c2;
SUBC> cook c3;
SUBC> dfit c4.
```

This command puts Cook's distances into column c3 and the DFFITS into column c4.

Notice that the subcommand for DFFITS is written `dfit`. You should print columns c3 and c4 and check these results against the corresponding results in Exhibit 5.4.1.

Problems

M5.4.1 Use the data below and exhibit the appropriate Minitab commands to obtain the quantities required in (a)–(g).

ROW	Y	X
1	8.8	1.0
2	3.1	1.5
3	8.5	2.0
4	15.3	2.5
5	15.2	3.0
6	16.4	3.5
7	21.5	4.0
8	22.5	4.5
9	22.3	5.0
10	52.2	11.0

(a) $h_{4,4}$.

(b) DFFITS_2.

(c) Cook's distance c_9.

(d) $\hat{Y}(4.5)$.

(e) r_6.

(f) $\hat{e}_2$.

(g) Studentized deleted residual T_7.

5.5 Ill Conditioning and Multicollinearity

As discussed in Section 5.5 in the textbook, if the columns of the X matrix are approximately linearly related, then multicollinearity exists and it may be very difficult to obtain reliable estimates of the β_i. Several diagnostic measures have been suggested for detecting the presence of approximate linear relationships among the predictor variables. One of the measures is the **variance inflation factor**. In Minitab, this quantity can be computed for each factor X_j using the subcommand `vif` of the `regress` command. To illustrate, we use the insurance data in Table 5.2.1. These data are stored in the file **premiums.mtp** on the data disk. You should first retrieve these data. The Minitab command and subcommand to obtain the variance inflation factors are

COMMAND FOR COMPUTING VARIANCE INFLATION FACTORS

```
MTB > regress c1 2 c2 c3;
SUBC> vif.
```

Minitab responds with:

```
--------------------------------------------------------------------

The regression equation is
Premium = 6.9 - 5.10 age + 0.0395 price

Predictor        Coef        Stdev     t-ratio         p        VIF
Constant         6.90        24.52        0.28     0.780
age           -5.0996       0.3894      -13.10     0.000        1.0
price        0.039533     0.001209       32.69     0.000        1.0

s = 41.94        R-sq = 97.7%      R-sq(adj) = 97.6%

Analysis of Variance

SOURCE         DF          SS          MS          F         p
Regression      2     2492087     1246043     708.49     0.000
Error          33       58038        1759
Total          35     2550125
--------------------------------------------------------------------
```

The column `VIF` in the output gives the value of the variance inflation factor for each predictor variable. For these data the value of VIF is 1.0 for each predictor and so there is no indication of multicollinearity.

Chapter 6

Applications of Regression I

6.1 Overview

There are no calculations in this section that require Minitab.

6.2 Prediction Intervals

In this section we explain a macro for calculating predicted values and prediction intervals for the mean and the sum of h future Y values. The name of the macro is **pred**, and the Minitab commands for the macro are stored in the files **pred.mtb**, **pred0.mtb**, and **pred1.mtb** in the subdirectory macro on the data disk. Note that predicted values and prediction intervals for a single future Y value can be obtained using this macro by taking $h = 1$.

To explain the macro we use part (2) of Task 6.2.1 where an agency evaluates the performance of used cars. It wants to obtain a 95% two-sided prediction interval for

$$Y_1(6.0, 24, 48.9) + Y_2(15.0, 21, 32.1)$$

which is the *sum* of the first-year maintenance costs for two cars, where car 1 will be driven 6,000 miles the first year after it is purchased, is 24 months old, and has 48,900 miles registered on its odometer, whereas car 2 will be driven 15,000 miles the first year after it is purchased, is 21 months old, and has 32,100 miles registered on its odometer. The data are given in Table 6.2.1 and are stored in the file **usedcars.mtp** on the data disk.

To use the macro, you must have the response variable Y in column c1 and the predictor variables $X_1, X_2, \ldots, X_k$ in consecutive columns beginning with c2, in this order. At most nine predictor variables are allowed in this macro. The values of the predictor variables for the h future Y values of interest are in the vectors x_1, x_2, ..., x_h where

$$x_1 = \begin{bmatrix} 1 \\ x_{1,1} \\ x_{1,2} \\ \vdots \\ x_{1,k} \end{bmatrix} \quad x_2 = \begin{bmatrix} 1 \\ x_{2,1} \\ x_{2,2} \\ \vdots \\ x_{2,k} \end{bmatrix} \quad \cdots \quad x_h = \begin{bmatrix} 1 \\ x_{h,1} \\ x_{h,2} \\ \vdots \\ x_{h,k} \end{bmatrix}$$

when an intercept term is present in the model (see (6.2.6)), but

$$x_1 = \begin{bmatrix} x_{1,1} \\ x_{1,2} \\ \vdots \\ x_{1,k} \end{bmatrix} \quad x_2 = \begin{bmatrix} x_{2,1} \\ x_{2,2} \\ \vdots \\ x_{2,k} \end{bmatrix} \quad \cdots \quad x_h = \begin{bmatrix} x_{h,1} \\ x_{h,2} \\ \vdots \\ x_{h,k} \end{bmatrix}$$

when there is no intercept term in the model. These vectors must be in h consecutive columns beginning with c12. They may be entered into the Minitab worksheet using the `set` command. Thus, for the current problem, retrieve the data from the file **usedcars.mtp** and then enter the vectors x_1 and x_2 in columns c12 and c13, respectively. Thus we have

$$\text{c12} = x_1 = \begin{bmatrix} 1 \\ 6.0 \\ 24.0 \\ 48.9 \end{bmatrix} \quad \text{and} \quad \text{c13} = x_2 = \begin{bmatrix} 1 \\ 15.0 \\ 21.0 \\ 32.1 \end{bmatrix}$$

Before using the macro, print the data and examine them to be sure that they are in the proper columns by using the command

```
print c1 c2 c3 c4 c12 c13
```

Minitab responds with:

ROW	mtcost	miles	age	odometer	C12	C13
1	190	6	70	70.7	1.0	1.0
2	379	11	72	70.9	6.0	15.0
3	201	6	98	108.8	24.0	21.0
4	194	8	60	49.0	48.9	32.1
5	189	4	84	95.7		
6	379	8	84	96.3		
7	183	6	64	64.6		
8	186	7	64	64.8		
9	456	19	60	49.2		
10	149	17	36	17.8		
11	175	5	66	68.1		
12	276	8	72	71.1		
13	277	8	72	70.9		
14	243	7	72	70.8		
15	195	7	66	68.3		
16	179	10	58	52.9		
17	186	8	61	58.8		
18	161	9	54	43.3		
19	267	11	60	49.7		
20	216	9	60	48.8		
21	130	15	36	18.5		
22	167	7	58	52.1		
23	211	18	48	31.7		
24	186	7	24	59.1		
25	165	9	54	43.6		
26	168	9	54	43.6		
27	148	6	60	49.3		
28	179	7	61	58.4		
29	186	7	64	64.6		
30	116	13	36	18.5		
31	312	9	72	71.1		
32	168	6	61	58.1		
33	190	7	28	49.7		
34	235	5	84	95.6		
35	195	8	29	48.3		

```
36    140     9    48      32.2
37    209     9    64      68.4
38    607    14    72      95.8
39    181     5    66      68.6
40    176     7    61      58.1
41    201    10    61      58.8
42    279     8    72      70.3
--------------------------------------------------------------------
```

The command to start the macro is

COMMAND TO START THE PREDICTION INTERVAL MACRO

```
MTB > execute 'b:\macro\pred'
```

Note: During the execution of the macro, Minitab will periodically pause, and issue the prompt `CONTINUE?` or the prompt `CONTINUE? (y for YES)` so that you will have ample time to read the output on the display. When you are ready to continue with the rest of the output, simply type `y` and press `Enter` .

Minitab responds with:

```
--------------------------------------------------------------------

This program will do calculations necessary to compute prediction
intervals for the sum and the average of h future Y values where
h is at most equal to 9.

The Y data must be in column c1 and the X data must be in
consecutive columns beginning with column  c2.  You will be asked
to input k, the number of predictors, and the desired confidence
coefficient. You will also be asked to input whether or not an
intercept term is present in the model.

The x vectors x1, x2,...,xh for which predictions are required
must be in consecutive columns beginning with column c12.
```

```
See (6.2.6). The first element of each vector is 1 when an
intercept term is present in the model. For example, suppose h=2
and the number of predictors is 4. Suppose the values of the
predictor variables for which predictions are required are

   X1 = 5.7    X2 = 12.0    X3 = 8.4    X4 = 11.5

for the first future Y value and

   X1 = 8.1    X2 =  3.9    X3 = 5.3    X4 = 13.1

for the second future Y value. The x1 and x2 vectors in c12 and
c13 are

                        c12   c13

                          1     1
                        5.7   8.1
                       12.0   3.9
                        8.4   5.3
                       11.5  13.1

if the model has an intercept term. If the model does not have
an intercept term, then the x1 and x2 vectors in c12 and c13 are

                        c12   c13

                        5.7   8.1
                       12.0   3.9
                        8.4   5.3
                       11.5  13.1

*******************************************************************

 Input the value of k, the number of predictor variables (note
 that k must be less than or equal to 9)
DATA> 3

 Does your model have an intercept term? If yes, enter 1. If no,
 enter 0.
DATA> 1
```

```
  Input h, the number of future Y values under consideration.
DATA> 2

  Input the value of the confidence coefficient 1-alpha
DATA> .95

=====================================================================

  The estimate of the average, Y_A, of the h future Y values is

  YAhat =      207.8350


  The standard error of this estimate is     43.68084


  For a two-sided prediction interval for the average Y_A of the
  h future Y values, with confidence coefficient  0.950
  the lower bound is      119.4078 and
  the upper bound is      296.2623

  The estimate of the sum, Y_S, of the h future Y values is

  YShat =      415.6700


  The standard error of this estimate is     87.36169


  For a two-sided prediction interval for the sum Y_S of the
  h future Y values, with confidence coefficient  0.950
  the lower bound is      238.8155 and
  the upper bound is      592.5246
---------------------------------------------------------------------
```

Thus the estimate of Y_A, the average of the first-year maintenance costs of these two cars, is $\hat{Y}_A$ =\$207.84. The standard error of $\hat{Y}_A$ is

$$SE(\hat{Y}_A) = \$43.6808$$

The confidence statement for a 95% prediction interval for Y_A is

$$C[\$119.41 \leq Y_A \leq \$296.26] = 0.95$$

The macro also gives the point estimate, the standard error, and a two-sided prediction interval for Y_S, the sum of the first-year maintenance costs of the two cars. In particular, we have

$$\hat{Y}_S = \$415.67, \quad SE(\hat{Y}_S) = \$87.3617$$

and

$$C[\$238.82 \leq Y_S \leq \$592.53] = 0.95$$

Problems

M6.2.1 (a) For the used cars data in Task 6.2.1 which is also in the file **usedcars.mtp**, find the point estimate of the first-year maintenance cost of each of three cars chosen at random from the following subpopulations:

(i) Car 1 will be driven 10,000 miles, is 20 months old, and has 30,200 miles on its odometer.

(ii) Car 2 will be driven 8,500 miles, is 15 months old, and has 15,000 miles on its odometer.

(iii) Car 3 will be driven 6,500 miles, is 24 months old, and has 28,000 miles on its odometer.

(b) In (a) find a 90% prediction interval for the first-year maintenance cost for Car 1.

(c) In (a) find a point estimate and a 90% prediction interval for the total first-year maintenance cost of the three cars.

M6.2.2 Work Exercise 6.9.1 (c), (d), and (e), using the macro provided in this section.

6.3 Tolerance Intervals

In this section we explain how to use a macro we have written for computing tolerance intervals discussed in Section 6.3. We illustrate by using parts (a) and (b) of Example 6.3.3 and computing a point estimate and a two-sided 95% confidence interval for the tolerance point $\lambda_{0.80}(3.0)$. The macro is named **toleranc** and the commands for this macro are stored in the files **toleranc.mtb**, **toleran0.mtb**, and **toleran1.mtb** in the subdirectory macro on the data disk. The data are given in Table 6.3.1 and are also stored in the file **table631.mtp** on the data disk. You should first retrieve the data from this file.

To use the macro, you must have the Y data in column c1 and the data for predictors $X_1, X_2, \ldots, X_k$ in consecutive columns beginning with column c2. The number of predictor variables k must be less than or equal to 9. The macro will ask you to input k, and also the quantities in (1), (2), (3), and (8) in Box 6.3.1. The command to start the macro is

COMMAND TO START THE TOLERANCE INTERVAL MACRO

```
MTB > execute 'b:\macro\toleranc'
```

Minitab responds with:

```
------------------------------------------------------------------------

This macro will do the calculations necessary to obtain estimates
for tolerance points. It will also calculate confidence intervals
for the tolerance points.

The Y data must be in column c1.

The X data must be in consecutive columns beginning with c2.

The number of predictors k must be less than or equal to 9.

You must be ready to input the following information:

    (a) The value of k, the number of predictors in the model.
    (b) Whether or not an intercept term is present in the model.
    (c) The x vector identifying the subpopulation for which a
        point estimate and a confidence interval for a tolerance
        point are required.
    (d) The value of p.
    (e) The value of 1-alpha, the desired confidence coefficient.
    (f) The noncentral t-value from Table T-8 in Appendix T
        corresponding to specified values of alpha/2, degrees
        of freedom and noncentrality parameter delta.
    (g) The noncentral t-value from Table T-8 in Appendix T
        corresponding to specified values of 1-alpha/2, degrees
        of freedom and noncentrality parameter delta.
```

```
**********************************************************************

  Input the value of k, the number of predictor variables (note k
  must be less than or equal to 9)
DATA> 1

  Does your model contain an intercept term? If yes, input 1. If no,
  input 0.
DATA> 1

  Input the  x-vector corresponding to the subpopulation for which
  a point estimate and a confidence interval for a tolerance point
  are required. Remember, if the model has an intercept term, then
  the x-vector contains k+1 elements, one for the intercept term and
  one for each predictor variable. In particular, the first element is
  the number 1 in this case. If the model does not contain an intercept
  term, then the x-vector contains k elements, one for each predictor
  variable.
DATA> 1 3

  Input the value of p.
DATA> .8

  Input the confidence coefficient 1-alpha for a two-sided tolerance
  interval.
DATA> .95

  Input the noncentral t-value in (8) of Box 6.3.1.
  This value is obtained from Table T-8.
  The value of alpha/2 is 0.02500.
  The degrees of freedom are    16 and
  the value of delta is    -3.241651

DATA> -6.174

  Input the noncentral t-value in (8) of Box 6.3.1.
  This value is obtained from Table T-8.
  The value of 1-alpha/2 is  0.9750.
  The degrees of freedom are    16 and
  the value of delta is    -3.241651

DATA> -1.254
```

```
==========================================================================

  The estimate of the tolerance point is    0.6624243


  A two-sided    95% confidence interval for the p-th tolerance
  point, where p =    0.800, has
  lower bound =     0.1179971    and
  upper bound =      1.465609.
--------------------------------------------------------------------------
```

Thus we see that the point estimate of $\lambda_{.80}(3.0)$ is 0.6624 and the 95% confidence statement for $\lambda_{.80}(3.0)$ is

$$C[0.1180 \le \lambda_{.80}(3.0) \le 1.4656] = 0.95$$

These agree with the results in Example 6.3.3 (except for rounding errors).

Problems

M6.3.1 In Example 6.3.3 find a point estimate and a 95% two-sided confidence interval for $\lambda_{0.20}(3.0)$, the number such that 20% of the subpopulation at $x = 3.0$ is below it. Use the macro **toleranc** discussed in this section.

M6.3.2 Work Exercise 6.9.2 using the macro discussed in this section.

6.4 Calibration and Regulation for Straight Line Regression

There is no command in Minitab that will compute point estimates and confidence regions for parameters in calibration and regulation problems, so we have written a macro that can be used for this purpose. To illustrate, we use Example 6.4.3 where the data are given in Table 6.4.1 and are stored in the file **thermom.mtp** on the data disk. The response variable Y is the temperature reading on a thermometer, and the predictor variable X is the true temperature.

The macro to compute the point estimate and confidence region for x_0, for a given value of y_0, is called **calib**. Minitab commands for this macro are stored in the files **calib.mtb, caliba.mtb, calibbc.mtb, calibb.mtb,** and **calibc.mtb** in the subdirectory macro on the data disk. You will be asked to input the number of the column that contains the Y values, the number of the column that contains the X values, the value of y_0, and the confidence coefficient. You should first retrieve the data from the file **thermom.mtp**. The command to start the macro is

COMMAND FOR CALIBRATION COMPUTATIONS

```
MTB > execute 'b:\macro\calib'
```

Minitab responds with:

```
-------------------------------------------------------------------------

  This macro will compute a point estimate and a confidence
  region for the point x0 in the calibration problem.

  Columns c80 through c100 should be left empty for workspace.

*******************************************************************

  Input the number of the column containing the  Y  values.
DATA> 1

  Input the number of the column containing the  X  values.
DATA> 2

  Input y0, the observed value of Y for which a point estimate and
          a confidence region for x0 are desired.
DATA> 104

  Input the desired confidence coefficient 1-alpha.
DATA> .95

===================================================================
```

```
The point estimate of x_0 is      103.9949


The required confidence region is in fact an interval and
it is given below.


For a two-sided confidence interval for x0
with confidence coefficient equal to  0.950

the lower bound is      103.4040  and
the upper bound is      104.5895
```

If the confidence region is not an interval, the macro will tell you so. In this problem we get $\hat{x}_0 = 103.995$, and the 95% confidence interval is given by

$$C[103.4040 \leq x_0 \leq 104.5895] = 0.95$$

Next we demonstrate another macro, supplied on the data disk, that will perform the computations required for obtaining a point estimate and a confidence region for x_0 in a **regulation** problem. The macro is named **regul** and the Minitab commands for this macro are in the files **regul.mtb, regula.mtb, regulbc.mtb** , **regulb.mtb,** and **regulc.mtb** in the subdirectory `macro` on the data disk. In Example M6.4.1 we illustrate the use of this macro.

Example M6.4.1

An investigator is studying the relationship of X, the amount of a new compound added to cement, and Y, the compression strength of the cement blocks. An experiment is conducted where specified amounts of the compound are added to the cement mixture and the strengths of the blocks are measured. We suppose that assumptions **(A)** are satisfied and that the data are obtained by sampling with preselected X values. The data are given below and are also stored in the file **cement.mtp**. The investigator wants to determine x_0, the amount of the compound that must be used, if the **average** strength of the population of cement blocks is to be 10,000 pounds per square inch. We use the **regul** macro to compute a point estimate and a 90% confidence region for x_0. You will be asked to input the number of the column that contains the Y values, the number of the column that contains the X values, the value of the mean `mu(x0)` (i.e., $\mu_Y(x_0)$), and the confidence coefficient.

Cement Data.

Y Strength (in thousands of lbs)	X Amount of Compound (in percent)
8.8	5
9.2	7
9.8	8
11.1	9
11.5	10
11.6	11
13.1	12
12.8	13
14.7	15
16.1	20

First retrieve the data from the file **cement.mtp**. The command to start the macro **regul** is

COMMAND FOR REGULATION COMPUTATIONS

```
MTB > execute 'b:\macro\regul'
```

Minitab responds with:

```
-------------------------------------------------------------------------

  This macro will compute a point estimate and a confidence
  region for the point x0 in the regulation problem.

  Columns c80 through c100 should be left empty for workspace.

**********************************************************************

  Input the number of the column containing the  Y  values.
DATA> 1
```

```
  Input the number of the column containing the  X  values.
DATA> 2


  Input mu(x0), the value of the subpopulation mean for which
  a point estimate and a confidence region for x0 are desired.
DATA> 10


  Input the desired confidence coefficient 1-alpha.
DATA> .90

================================================================

  The point estimate of x_0 is     7.497659

  The required confidence region is in fact an interval and
  it is given below.


  For a two-sided confidence interval for x0
  with confidence coefficient equal to  0.900

  the lower bound is     6.693408  and
  the upper bound is     8.171340
----------------------------------------------------------------
```

Thus we see that $\hat{x}_0 = 7.498$ and the 90% confidence statement is

$$C[6.6934 \le x_0 \le 8.1713] = 0.90$$

If the confidence region is not an interval, the macro will tell you so.

Problems

M6.4.1 Work Problems 6.4.1, 6.4.2, 6.4.3, 6.4.4, and 6.4.5 in the textbook using the macros discussed in this section.

M6.4.2 Work Example M6.4.1 if the investigator wants a point estimate and a 90% confidence region for x_0, the amount of the new compound to be added so that the average strength will be 15,000 pounds.

6.5 Comparison of Several Straight Line Regressions – Identical, Parallel, and Intersecting Lines

In this section we discuss a macro that we have written and supplied on the data disk that can be used to perform the computations for comparing several regression functions. Specifically, the program will calculate point estimates and simultaneous confidence intervals for m linear combinations of α_i and β_j with confidence coefficient $1-\alpha$. See (6.5.15). The macro is called **compare** and the Minitab commands for this macro are stored in the files **compare.mtb, compare1.mtb, compare2.mtb**, and **compare3.mtb** in the subdirectory macro on the data disk. We illustrate the macro by using Example 6.5.3, where the data are given in Table 6.5.2 and are also stored in the file **eggshell.mtp** on the data disk. We want simultaneous confidence intervals for the six (i.e., $m = 6$) linear combinations given below.

$$\alpha_1 - \alpha_2,\ \alpha_1 - \alpha_3,\ \alpha_2 - \alpha_3,\ \beta_1 - \beta_2,\ \beta_1 - \beta_3,\ \beta_2 - \beta_3$$

To use the macro, you must have the data organized as follows:

Y_1 must be in column c1
X_1 must be in column c2
Y_2 must be in column c3
X_2 must be in column c4
Y_3 must be in column c5
X_3 must be in column c6

and so forth, if more than 3 regression functions are being compared. Here, Y_i and X_i, $i = 1, 2, 3$, represent the response variable and the predictor variable, respectively, for the ith regression function. The macro will ask you to input the following quantities.

(1) h, the number of regression lines to be compared, which in this example is 3.

(2) m, the number of linear combinations of the parameters of interest, which in this example is 6.

(3) The simultaneous confidence coefficient, which in this example is 0.95.

(4) The coefficients for each linear combination of interest.

You should first retrieve the data from the file **eggshell.mtp**. The command to start the macro is

COMMAND FOR COMPARING REGRESSION FUNCTIONS

```
MTB > execute 'b:\macro\compare'
```

Minitab responds with:

```
--------------------------------------------------------------------------

  This macro will perform the calculations for comparing h
  (two or more) regression lines.

  Leave columns c80-c100 empty for workspace.

  Organize the data for the  h  regression lines to be compared
  so that they are in the following format.

  Y1,X1, Y2,X2, Y3,X3, ..., Yh,Xh

  beginning in column c1 and ending in column c(2h). For instance,
  if there are 3 regression lines, the data should occupy columns
  c1 through c6 in the order Y1,X1,Y2,X2,Y3,X3.

**********************************************************************

  Input the number of regression lines to be compared.
DATA> 3

  Input the number of intervals desired.
DATA> 6

  Input the simultaneous confidence coefficient desired.
DATA> .95

  Input the coefficients for the h linear functions of interest.
  A typical linear function will have the form

  a(1) alpha(1) + b(1) beta(1)+ a(2) alpha(2) + b(2) beta(2) +
  ... + ... + a(h) alpha(h) + b(h) beta(h)

  You should enter m rows of coefficients, one set of coefficients
  for each linear function.
```

Example:

Suppose there are h = 3 regression lines to be compared and m = 2 linear functions are of interest. Suppose these two linear functions are

```
alpha(1) + 10 beta(1) - alpha(3) -10 beta(3) and
alpha(2) + 15 beta(2) - alpha(3) -15 beta(3)
```

Then you should enter the coefficients in two rows in response to the Minitab prompt DATA> as follows:

```
DATA> 1 10 0 0 -1 -10
DATA> 0 0 1 15 -1 -15
```

Don't omit the zero coefficients.

Input the coefficients for linear function 1

```
DATA> 1 0 -1 0 0 0
```

Input the coefficients for linear function 2

```
DATA> 1 0 0 0 -1 0
```

Input the coefficients for linear function 3

```
DATA> 0 0 1 0 -1 0
```

Input the coefficients for linear function 4

```
DATA> 0 1 0 -1 0 0
```

Input the coefficients for linear function 5

```
DATA> 0 1 0 0 0 -1
```

Input the coefficients for linear function 6

```
DATA> 0 0 0 1 0 -1
```

```
==========================================================================

 The pooled estimate of the subpopulation standard deviation
 is      1.217584    and the associated degrees of freedom =   23


 The required simultaneous confidence intervals (Bonferroni)
 with simultaneous confidence coefficient greater than or
 equal to    0.950 are given below


 Linear Function    Point Estimate     Lower Bound     Upper Bound


         1             -0.5011616          -4.284904        3.282580
         2              0.9436436          -2.571795        4.459082
         3               1.444805          -2.379294        5.268904
         4               1.954605           1.452158        2.457051
         5               2.785993           2.440870        3.131115
         6              0.8313880          0.3708251        1.291951
--------------------------------------------------------------------------
```

Thus the point estimates are

$$\begin{aligned}
\hat{\alpha}_1 - \hat{\alpha}_2 &= -0.50116 \\
\hat{\alpha}_1 - \hat{\alpha}_3 &= 0.94364 \\
\hat{\alpha}_2 - \hat{\alpha}_3 &= 1.44481 \\
\hat{\beta}_1 - \hat{\beta}_2 &= 1.95460 \\
\hat{\beta}_1 - \hat{\beta}_3 &= 2.78599 \\
\hat{\beta}_2 - \hat{\beta}_3 &= 0.83139
\end{aligned}$$

We are at least 95% confident that the six confidence intervals below are simultaneously correct.

$$\begin{aligned}
-4.284904 \le \alpha_1 - \alpha_2 \le 3.282580 \\
-2.571795 \le \alpha_1 - \alpha_3 \le 4.459082 \\
-2.379294 \le \alpha_2 - \alpha_3 \le 5.268904 \\
1.452158 \le \beta_1 - \beta_2 \le 2.457051 \\
2.440870 \le \beta_1 - \beta_3 \le 3.131115 \\
0.3708251 \le \beta_2 - \beta_3 \le 1.291951
\end{aligned}$$

Problems

M6.5.1 In Example 6.5.3, use the macro discussed in this section to find confidence intervals for

$$\alpha_1 - \alpha_2, \quad \alpha_1 - \alpha_3, \quad \alpha_2 - \alpha_3$$

so that you have at least 90% confidence that all intervals are simultaneously correct.

6.6 Intersection of Two Straight Line Regression Functions

In this section we discuss the macro **inter** that can be used to compute a point estimate and a confidence region for x_0, the point where two straight line regression functions intersect. The macro commands are stored in the files **inter.mtb, inter1.mtb, intera.mtb, interb.mtb,** and **interc.mtb** in the subdirectory macro on the data disk. We illustrate this macro by using Example 6.6.2, where an investigator wants to compare the hardness of eggshells for breeds 2 and 3 for values of the food supplement in the range from 2 to 20 units. To help make this comparison we want to determine x_0, the point where the regression lines for breed 2 and breed 3 intersect. We obtain a point estimate of x_0 and a 95% confidence region for x_0. The data are given in Table 6.6.1 and are also stored in the file **eggshell.mtp** on the data disk. To use the macro for this example, you must first retrieve the data. During execution, the macro will ask you to input the column numbers for the columns containing the Y values and the X values for the two regression lines under study. You will also be asked to input $1-\alpha$, the confidence coefficient for a confidence region for x_0.

The command to start the macro is

COMMAND FOR COMPUTING THE POINT OF INTERSECTION OF TWO LINES

```
MTB > execute 'b:\macro\inter'
```

Minitab responds with:

```
-------------------------------------------------------------------------

  This macro will calculate a point estimate and a confidence region
  for the point of intersection of two straight line regression
  functions.

***********************************************************************

  Input the number of the column containing the Y values
  for the first regression line.
DATA> 3

  Input the number of the column containing the X values
  for the first regression line.
DATA> 4

  Input the number of the column containing the Y values
  for the second regression line.
DATA> 5

  Input the number of the column containing the X values
  for the second regression line.
DATA> 6

  Input the confidence coefficient desired.
DATA> .95
=======================================================================

  The point estimate for the point of intersection is    -1.737823


  The confidence region with confidence coefficient equal to  0.950
  is in fact an interval.

  The lower bound is    -7.620478  and
  the upper bound is     1.087612

-------------------------------------------------------------------------
```

Thus the required confidence statement is

$$C[-7.62 \leq x_0 \leq 1.09] = 0.95$$

Using this confidence interval we would perhaps conclude that the average hardness of eggshells will be greater for breed 2 than for breed 3 for all values of food supplement in the range from 2 to 20.

Problems

M6.6.1 For the eggshell data in Example 6.5.3, use the macro discussed in this section and find a point estimate and a 90% confidence region for x_0, the point where the straight line regression functions for breeds 1 and 3 intersect.

M6.6.2 Repeat Problem **M6.6.1** for breeds 1 and 2.

6.7 Maximum or Minimum of a Quadratic Regression Model

In this section we describe the macro **quadr** written by us, that can be used to compute a point estimate and a confidence region for x_0, the point where a quadratic regression function attains its maximum (or minimum) value. The macro commands are stored in the files **quadr.mtb**, **quadr1.mtb**, **quadra.mtb**, **quadrb.mtb**, and **quadrc.mtb** in the subdirectory `macro` on the data disk. The data must be organized so that the sample Y values are in column `c1` and the sample X values are in column `c2`. We use Example 6.7.3 to illustrate this macro. The data are given in Table 6.7.2 and are also stored in the file **sulfuric.mtp** on the data disk.

We will obtain a point estimate and a 95% confidence region for x_0, the temperature at which the process should be run so that the output will be maximized. First we retrieve the data from the file **sulfuric.mtp**. The command to start the macro is

COMMAND TO COMPUTE THE POINT OF MAXIMUM OR MINIMUM OF A QUADRATIC REGRESSION FUNCTION

```
MTB > execute 'b:\macro\quadr'
```

Minitab responds with:

```
------------------------------------------------------------------------

  This macro will do the calculations necessary to obtain
  a point estimate and a confidence region for the point
  where the MAXIMUM (or MINIMUM) of a QUADRATIC REGRESSION
  function occurs.

  The data should be organized so that
  Y is in column c1 and X is in column c2.

  Leave columns c80-c100 blank for workspace.

**********************************************************************

  Input the confidence coefficient desired.
DATA> .95

======================================================================

  The estimate of the point of extremum is     272.2905


  The confidence region with confidence coefficient equal to  0.950
  is in fact an interval.

  The lower bound is      257.4099  and
  the upper bound is      293.5309

------------------------------------------------------------------------
```

Thus the maximum yield is estimated to occur at 272.29° C. A 95% confidence statement for x_0, the temperature at which the maximum yield occurs, is

$$C[257.41 \leq x_0 \leq 293.53] = 0.95$$

Problems

M6.7.1 For Problem 6.7.1, use the macro discussed in this section to find a 90% confidence region for x_0, the amount of sand to use to maximize the crushing strength of the concrete cylinders.

6.8 Linear Splines

In this section we explain the macro **spline** which can be used to calculate point estimates and confidence intervals for spline regression functions. The macro commands are stored in the file **spline.mtb** in the subdirectory macro on the data disk. We illustrate the use of this macro by reworking Example 6.8.3. The data for this example are given in Table 6.8.1 and are also stored in the file **sales.mtp** on the data disk. Recall that, in this example the population (spline) regression function is given by

$$\mu_Y(x) = \begin{cases} \mu_Y^{(1)}(x) = \alpha_1 + \beta_1 x & \text{for } 0 \le x \le 50 \\ \\ \mu_Y^{(2)}(x) = \alpha_2 + \beta_2 x & \text{for } 50 \le x \le 100 \end{cases}$$

To use the macro you must have the data organized such that the sample Y values are in column c1 and the sample X values are in column c2. The macro will automatically print out the point estimates for α_1, β_1, α_2, β_2, and σ. In addition, you can compute one-at-a-time $1 - \alpha$ confidence intervals for as many linear functions

$$a_1\alpha_1 + b_1\beta_1 + a_2\alpha_2 + b_2\beta_2$$

as desired. From these you can obtain simultaneous confidence intervals by the Bonferroni method if you want. You must input the knot point q, the desired confidence coefficient for one-at-a-time two-sided confidence intervals, and the coefficients a_1, a_2, b_1, b_2 of the linear functions of interest, when requested by the program.

To illustrate the use of the macro **spline**, we obtain point estimates and 95% confidence intervals for five linear functions, viz., α_1, β_1, α_2, β_2, and $\mu_Y(65) = \alpha_2 + 65\beta_2$, in Example 6.8.3 (since $65 > q = 50$, we use the second regression line to evaluate $\mu_Y(65)$). First we retrieve the data from the file **sales.mtp**. The command to start the macro is

COMMAND FOR LINEAR SPLINE MACRO

```
MTB > execute 'b:\macro\spline'
```

Minitab responds with:

```
----------------------------------------------------------------------

  This macro will calculate point estimates for the parameters in
  a linear spline model with a single known knot point. It will also
  calculate confidence intervals for specified linear functions of
  the parameters.

  Y data should be in column c1 and X data in column c2.

  Keep columns c80-c100 empty for workspace.

*********************************************************************

  Input the knot point q.
DATA> 50

  Input the number of linear functions for which confidence
  intervals are required.
DATA> 5

  Input the coefficients for the linear functions of interest.
  Note that each linear function of interest can be written in
  the form

    a(1) alpha(1) + b(1) beta(1) + a(2) alpha(2) + b(2) beta(2).

  The coefficients to be input are

    a(1)  b(1)  a(2)  b(2)

  in this order.

DATA> 1 0 0 0
DATA> 0 1 0 0
DATA> 0 0 1 0
DATA> 0 0 0 1
DATA> 0 0 1 65
     5 ROWS READ

  Input the confidence coefficient for two-sided one-at-a-time
  confidence intervals for each linear function.
DATA> .95
```

```
=====================================================================

Point estimates for the intercepts and slopes of the two lines
are given below. Line 1 is the line to the left of the knot point
and line 2 is to the right of the knot point.

Estimated intercept for line 1 is      201.4454
Estimated   slope   for line 1 is      5.021784

Estimated intercept for line 2 is      404.2462
Estimated   slope   for line 2 is     0.9657681


Estimated subpopulation standard deviation is       11.04885

The point estimates and corresponding standard errors for the
linear functions of interest are given below. Also given are
lower bounds and upper bounds for two-sided confidence intervals
with one-at-a-time confidence coefficient equal to  0.950

 Linear          Point      Standard          Lower          Upper
Function      Estimate         Error          Bound          Bound


        1     201.4454      11.69917       176.7622       226.1285
        2     5.021784     0.2874708       4.415273       5.628296
        3     404.2462      15.27978       372.0086       436.4838
        4    0.9657681     0.2398292      0.4597715       1.471765
        5     467.0211      3.130059       460.4173       473.6250
---------------------------------------------------------------------
```

Thus we see that the point estimates and 95% confidence intervals for α_1, β_1, α_2, β_2, and $\mu_Y(65)$ are as follows:

Parameter	Point estimate	Confidence interval
α_1	201.4454	$176.7622 \le \alpha_1 \le 226.1285$
β_1	5.0218	$4.4153 \le \beta_1 \le 5.6283$
α_2	404.2462	$372.0086 \le \alpha_2 \le 436.4838$
β_2	0.9658	$0.4598 \le \beta_2 \le 1.4718$
$\mu_Y(65)$	467.0211	$460.4173 \le \mu_Y(65) \le 473.6250$

Note that, since $65 > q = 50$, we have $\mu_Y(65) = \alpha_2 + 65\beta_2$. Consequently, we enter the coefficients `0 0 1 65` to find a point estimate and a confidence interval for $\mu_Y(65)$.

Problems

M6.8.1 This problem refers to the data and the model discussed in Example 6.8.3, where Y is sales and X is money spent on advertising. In this problem $q = 50$.

(a) Find point estimates of α_1, β_1, α_2, and β_2.

(b) Find one-at-a-time 90% confidence intervals for the quantities in (a).

(c) Plot the fitted values $\hat{\mu}_Y(x_i)$ against x_i.

(d) Compute a point estimate and a 90% confidence interval for $\mu_Y(75)$, the average sales (in thousands of dollars) a company expects if it plans to spend 75 thousand dollars on advertising.

Chapter 7

Applications of Regression II

7.1 Overview

No computing instructions are needed in this section.

7.2 Subset Analysis and Variable Selection

No computing instructions are needed in this section.

7.3 All-Subsets Regression

In this section and the next, we illustrate Minitab commands that can be used for variable selection. There is no single command in the present version of Minitab that will give a complete set of results for **all-subsets** regression. However, the Minitab command `breg` (which stands for best subsets `r`egression) will carry out a **best subsets** analysis. This analysis is, in principle, an **all-possible-subsets** analysis that concentrates on finding only the **best few** subset models of each size.

Using the **breg** command in Minitab, one can obtain the values of s, R^2, *adj*-R^2, and C_p for the few best models (the user can specify the number of best models to be

1, 2, 3, 4, or 5 – the default value is 2) for each possible number of predictors. For example, in a problem where the total number of predictors is 4 ($k = 4$), Minitab will display the values of s, R^2, adj-R^2, and C_p for the following.

(1) The two best models when one predictor is used.

(2) The two best models when two predictors are used.

(3) The two best models when three predictors are used.

(4) The best model when all four predictor variables are used (there is only one model in this case, and it is the model using all four predictor variables).

In the above, instead of the two best models for each subset size, the user can specify any of the following: the one best model, the two best models, the three best models, the four best models, or the five best models, with the subcommand `best k` after the `breg` command.

As an illustration, we use the GPA data of Example 4.4.2 that are given in Table 4.4.3 and in the file **gpa.mtp** on the data disk. The data for the response variable Y are in column `c1`, and the data for the predictor variables X_1, X_2, X_3, and X_4 are in columns `c2`, `c3`, `c4`, and `c5`, respectively. We use the command `breg` (best subsets regression) to find the best few models for this example for each possible number of predictors. In the `breg` command, the first column specified, namely `c1` in the command below, must contain the values of the response variable Y, and the subsequent columns, `c2 c3 c4 c5` (or simply `c2-c5`) in the command below must contain the values of the predictor variables. First retrieve the file **gpa.mtp** and then issue the following command.

COMMAND FOR BEST SUBSETS REGRESSION

```
MTB > breg c1 c2-c5
```

Minitab responds with:

```
-------------------------------------------------------------------

Best Subsets Regression of GPA

                                                     S S
                                                     A A H H
                                                     T T S S
                                                     m v m e
                                                     a e a n
                  Adj.                               t r t g
Vars     R-sq     R-sq       C-p            s        h b h l

   1     72.2     70.6      12.4      0.33700        X
   1     48.3     45.4      36.7      0.45949            X
   2     81.1     78.9       5.3      0.28578        X X
   2     79.2     76.7       7.2      0.29982        X   X
   3     85.0     82.2       3.2      0.26211        X X X
   3     81.1     77.6       7.3      0.29454        X X   X
   4     85.3     81.4       5.0      0.26851        X X X X
-------------------------------------------------------------------
```

This computer output contains the following information. When considering subset models with a **single** predictor, the best single variable is SATmath (using any of the four criterion functions s, R^2, $adj\text{-}R^2$, or C_p). Look across the line that contains the first 1 in the column labeled Vars (variables) – where the 1 stands for a model with a single predictor – until you come to the symbol X. Notice that X is below SATmath, so SATmath is the best single predictor for predicting Y. The value of R^2, $adj\text{-}R^2$, C_p, and s for the subset model containing SATmath only are

$$R^2 = 72.2\%, \quad adj\text{-}R^2 = 70.6\%, \quad C_p = 12.4, \quad \text{and} \quad s = 0.33700$$

The second best variable for predicting Y when there is only one predictor is obtained by looking across the line that contains the second 1 under the column labeled Vars until you come to the symbol X. This X is under HSmath, so HSmath is the second best single variable for predicting Y.

For best subset models that contain two predictors, look under the column labeled Vars until you come to the first 2 (the 2 stands for models with two predictor variables). Look across that line until you come to two X's. These X's are under SATmath and SATverb, so SATmath and SATverb lead to the best model among the subset models that contain exactly two predictors. You can look at the corresponding values of R^2, adj-R^2, C_p, and s to see how good this model is for predicting Y.

Since we did not specify the number of best subset models we wanted corresponding to each possible number of predictors, Minitab, by default, prints summary information for the best **two** subset models for each possible subset size. Of course, there will be only one model when all the predictors are included. From the computer output above, we see that the best one-variable model contains SATmath; the second best one-variable model contains HSmath. The best two-variable model contains SATmath and SATverb; the second best model with two variables contains SATmath and HSmath. The best three-variable model contains SATmath, SATverb, and HSmath. The second best three-variable model contains SATmath, SATverb, and HSengl. The best four-variable model contains SATmath, SATverb, HSmath, and HSengl. Of course there is only one four-variable model since the total number of predictors is four.

If you want the summary information to be printed for k best models for each possible subset size, you can use the subcommand

```
best k
```

where k must be an integer between 1 and 5. The command and subcommand for obtaining the best k models for each possible subset size in the GPA example are:

MINITAB COMMAND FOR BEST k MODELS OF EACH SUBSET SIZE

```
MTB > breg c1 c2-c5;
SUBC> best k.
```

In particular, for the GPA data, the best five models for each possible subset size can be obtained using the command and subcommand given below.

```
breg c1 c2-c5;
best 5.
```

Minitab responds with:

```
------------------------------------------------------------------------

Best Subsets Regression of GPA

                                                        S S
                                                        A A H H
                                                        T T S S
                                                        m v m e
                                                        a e a n
                    Adj.                                t r t g
Vars     R-sq       R-sq       C-p            s         h b h l

   1     72.2       70.6       12.4     0.33700         X
   1     48.3       45.4       36.7     0.45949             X
   1     42.7       39.5       42.4     0.48367           X
   1     36.8       33.3       48.4     0.50793               X
   2     81.1       78.9        5.3     0.28578         X X
   2     79.2       76.7        7.2     0.29982         X   X
   2     72.5       69.3       14.0     0.34470         X     X
   2     67.1       63.2       19.5     0.37714             X X
   2     63.5       59.2       23.2     0.39714           X X
   3     85.0       82.2        3.2     0.26211         X X X
   3     81.1       77.6        7.3     0.29454         X X   X
   3     80.2       76.5        8.2     0.30137         X   X X
   3     73.7       68.7       14.8     0.34771           X X X
   4     85.3       81.4        5.0     0.26851         X X X X
------------------------------------------------------------------------
```

As stated above, with the current version of Minitab it is not possible to have a table of measures such as this printed for all the possible subset models. The subcommand `best 5` results in the most detailed output possible. For the GPA data, there are six possible subset models that use two predictors, but in the computer output above we only have the best five models with two predictors. The model using SATverb and HSengl is not part of the output because it is not one of the best five models using two predictors. Of course, if necessary, we can obtain the values of R^2, *adj*-R^2, and s for this model by using the `regress` command and we can obtain the value of C_p for this model by performing a few additional calculations.

7.4 Alternative Methods for Subset Selection

In this section we discuss several commands and subcommands in Minitab that can be used to do the computations for backward, forward, and stepwise regression. We begin with stepwise regression since the backward and forward procedures are obtained from the stepwise procedure by using appropriate values for *F-in* and *F-out.*

Stepwise regression

The principal command for stepwise regression is the command `stepwise` which has as its arguments a number of columns. The first column specified designates the response variable Y, and all the succeeding columns designate the predictor variables $X_1, X_2, \ldots$. For instance, in the following command

```
stepwise c1 c2-c6
```

the response variable Y is in column `c1` and the predictor variables are in columns `c2` through `c6`. Of course, the response variable and the predictor variables can be in any columns, but the proper columns must be specified in the command.

FENTER and FREMOVE subcommands

Minitab has many subcommands that can be used with the `stepwise` command. As we discussed in Section 7.4 in the textbook, the stepwise procedure requires two numbers, *F-in* and *F-out.* In Minitab, the command to specify *F-in* is `fenter` and the command for specifying *F-out* is `fremove` . If the value for *F-in* is not specified by the user with the `fenter` subcommand, then Minitab uses the default value of 4.0. Likewise, if the value for *F-out* is not specified by the user with the `fremove` subcommand, then Minitab uses the default value of 4.0.

The stepwise procedure may consist of several steps depending on the number of predictor variables and the values of `fenter` and `fremove`. When the number of steps is too large to fit on one "page," Minitab splits the output into a number of "pages". The number of steps printed on each "page" depends on the width of your output device, but you can specify the number of steps per page using the subcommand `steps = k` . For details on this you should consult the *Minitab Reference Manual.* At the end of each "page" of output, Minitab prints the following:

```
MORE? (YES, NO, SUBCOMMAND, OR HELP)
```

```
SUBC>
```

If you want more "steps" printed, your response must be **yes** after the prompt SUBC> (then, as usual, press Enter), and if you do not want more "steps" printed, your response is no. Always answer **yes**. If the stepwise procedure has no more steps to calculate, then Minitab will respond with

```
 NO VARIABLES ENTERED OR REMOVED
 MORE? (YES, NO, SUBCOMMAND, OR HELP)
SUBC>
```

At this point you can respond with a no. Also, instead of **yes** or no, you can intervene at this stage by removing and/or adding some variables using appropriate subcommands. This variation of the stepwise procedure was not discussed in the textbook, so it is not discussed here.

The basic command and subcommands for stepwise regression have the following form (assume that the response variable is in column c1 and the predictor variables are in columns c2 to c6, and that you want to use **fenter** equal to 3 and **fremove** equal to 2).

STEPWISE REGRESSION COMMAND AND SUBCOMMANDS

```
MTB > stepwise c1 c2-c6;
SUBC> fenter=3.0;
SUBC> fremove=2.0.
```

We illustrate these using Example 7.4.3. The data are given in Table 4.4.3 and are also stored in the file **gpa.mtp** on the data disk. Retrieve the data from this file. The response variable Y is in column c1, and the predictor variables are in columns c2-c5. We now ask Minitab to perform a stepwise regression analysis using **fenter** equal to 4 and **fremove** equal to 3.

```
stepwise c1 c2-c5;
fenter=4.0;
fremove=3.0.
```

Minitab responds with:

```
--------------------------------------------------------------------------

 STEPWISE REGRESSION OF    GPA     ON  4 PREDICTORS, WITH N =   20

    STEP          1         2         3
CONSTANT    0.9670    0.5071    0.3342

SATmath    0.00318   0.00261   0.00218
T-RATIO       6.83      5.88      4.80

SATverb              0.00157   0.00131
T-RATIO                 2.83      2.50

HSmath                           0.180
T-RATIO                           2.05

S            0.337     0.286     0.262
R-SQ         72.17     81.10     85.04
 MORE? (YES, NO, SUBCOMMAND, OR HELP)
SUBC> yes

 NO VARIABLES ENTERED OR REMOVED

 MORE? (YES, NO, SUBCOMMAND, OR HELP)
SUBC> no
--------------------------------------------------------------------------
```

We notice that in step 1 the model is

$$\beta_0 + \beta_1 x_1 \quad (\text{i.e., } \beta_0 + \beta_1 \text{ SATmath})$$

and $\sqrt{MSE}$ for that model is denoted by `S` and is 0.337. Also, the model in step 1 has `R-SQ=72.17%`. The model in step 2 is

$$\beta_0 + \beta_1 x_1 + \beta_2 x_2 \quad (\text{i.e., } \beta_0 + \beta_1 \text{ SATmath } + \beta_2 \text{ SATverb})$$

with `S=0.286` and `R-SQ=81.10%`. The final model (which, in this case, is the model in step 3) is

$$\beta_0 + \beta_1 x_1 + \beta_1 x_2 + \beta_2 x_3 \quad (\text{i.e., } \beta_0 + \beta_1 \text{ SATmath } + \beta_2 \text{ SATverb } + \beta_3 \text{ HSmath})$$

with `S=0.262` and `R-SQ=85.04%`.

The final model is the same as the final model in Example 7.4.3, which is given in (7.4.61). The value of F_C after the first variable has been entered (which is variable X_1 = SATmath) is 46.67, given in (7.4.56). In the Minitab output a T-RATIO is given instead of F_C, and it is 6.83. But we know that if we square a t-statistic with, say, q degrees of freedom, it is the same as an F-statistic with 1 and q degrees of freedom. Thus $(6.83)^2 = 46.649$, which is the same as in (7.4.56) except for rounding errors.

The Subcommand ENTER

In Example 7.4.4, we again performed a stepwise regression using the GPA data with the same values of `fenter` and `fremove` as in Example 7.4.3, but with a different initial model. The initial model is

$$\beta_0 + \beta_3 x_3 + \beta_4 x_4$$

Thus the initial model contains the variables X_3 = HSmath and X_4 = HSengl. There is a subcommand in Minitab for specifying an initial model (different from the model containing only β_0) in stepwise regression. This subcommand is `enter` followed by the columns that contain the predictor variables that you want to include in the initial model. For this example it is

```
enter c4 c5
```

where `c4, c5` are the columns that contain the variables X_3, X_4 that are to be used in the initial model. When no initial model is specified using the `enter` subcommand, Minitab starts with the model β_0 as the initial model.

As an illustration, the command and subcommands for carrying out a stepwise regression analysis for the GPA data using $\beta_0 + \beta_3 x_3 + \beta_4 x_4$ as the initial model, `fenter` equal to 4, and `fremove` equal to 3, are

COMMAND FOR STEPWISE REGRESSION WITH USER SPECIFIED INITIAL MODEL

```
MTB > stepwise c1 c2 c3 c4 c5;
SUBC> enter c4 c5;
SUBC> fenter=4.0;
SUBC> fremove=3.0.
```

Minitab responds with:

```
---------------------------------------------------------------------

 STEPWISE REGRESSION OF   GPA    ON  4 PREDICTORS, WITH N =   20

    STEP        1        2        3        4
CONSTANT  -0.3400   0.2690   0.6438   0.3342

HSmath      0.417    0.247    0.233    0.180
T-RATIO      3.96     2.50     2.40     2.05

HSenglsh     0.58     0.18
T-RATIO      3.12     0.91

SATmath           0.00213  0.00251  0.00218
T-RATIO              3.26     5.03     4.80

SATverb                             0.00131
T-RATIO                                2.50

S           0.377    0.301    0.300    0.262
R-SQ        67.08    80.22    79.20    85.04
 MORE? (YES, NO, SUBCOMMAND, OR HELP)
SUBC> yes

 NO VARIABLES ENTERED OR REMOVED

 MORE? (YES, NO, SUBCOMMAND, OR HELP)
SUBC> no
---------------------------------------------------------------------
```

We see that the final model contains X_1, X_2, and X_3, which are SATmath, SATverb, and HSmath, respectively. The result is the same as in (7.4.75).

Next we explain how the `stepwise` command can be used to carry out a forward selection analysis or a backward elimination analysis.

Forward Selection

The Minitab command and subcommands for the forward selection procedure using `fenter=4.0` (assuming that the response variable is in column `c1` and the predictor variables are in consecutive columns `c2` to `c6`) are

COMMAND FOR FORWARD SELECTION PROCEDURE

```
MTB > stepwise c1 c2-c6;
SUBC> fenter=4.0;
SUBC> fremove=0.
```

You must, of course, specify the proper columns in the `stepwise` command for each problem. The value for `fenter` does not have to be 4.0; it can be any positive value of your choice. The value of `fremove` is set equal to zero to prevent the stepwise procedure from ever performing a backward step.

Backward Elimination

The Minitab command and subcommands for the backward elimination procedure, using `fremove` equal to 4 (assuming that the response variable is in column `c1` and the predictors are in consecutive columns `c2` to `c6`) are

COMMAND FOR THE BACKWARD ELIMINATION PROCEDURE

```
MTB > stepwise c1 c2-c6;
SUBC> fenter=100000;
SUBC> fremove=4.0;
SUBC> enter c2-c6.
```

You must, of course, specify the proper columns in the `stepwise` command for each problem. The value for `fremove` does not have to be 4.0; it can be any value. The reason that `fenter` is given the value 100,000 is that the value of `fenter` must be large enough so no variable will qualify for entering the model in the forward stage of the stepwise procedure.

Problems

M7.4.1

(a) Use the Minitab commands and subcommands discussed in this section to work Example 7.4.5.
(b) Repeat (a) using the forward selection procedure instead of the stepwise regression procedure. Use `fenter = 3.`
(c) Repeat (a) using the backward elimination procedure instead of the stepwise regression procedure. Use `fremove = 3.`

7.5 Growth Curves

In this section we discuss a Minitab macro that will enable you to do the computations necessary for the growth curve analysis as discussed in Section 7.5. This macro is named **growth**, and it will compute point estimates and confidence intervals for any specified linear combination

$$\theta = a^T\beta = a_0\beta_0 + a_1\beta_1 + \cdots + a_k\beta_k$$

of the β_i. The Minitab commands for this macro are stored in the file **growth.mtb** in the subdirectory macro on the data disk. To use this macro you must have the Y data organized in columns as in Table C below. Also see Table D and Table E, which are the same as Table 7.5.2 and Table 7.5.3, respectively, in the textbook.

Table C
A Schematic Representation of the Sample Data for a Growth Curve Study

Item	Response at time t_1	$\cdots$	Response at time t_j	$\cdots$	Response at time t_k
1	$y_{1,1}$	$\cdots$	$y_{1,j}$	$\cdots$	$y_{1,k}$
2	$y_{2,1}$	$\cdots$	$y_{2,j}$	$\cdots$	$y_{2,k}$
$\vdots$	$\vdots$	$\vdots$	$\vdots$	$\vdots$	$\vdots$
i	$y_{i,1}$	$\cdots$	$y_{i,j}$	$\cdots$	$y_{i,k}$
$\vdots$	$\vdots$	$\vdots$	$\vdots$	$\vdots$	$\vdots$
m	$y_{m,1}$	$\cdots$	$y_{m,j}$	$\cdots$	$y_{m,k}$

Table D
Drug Concentration Data (in milligrams/liter)

Subject	t_1 (1 hour)	t_2 (2 hours)	t_3 (3 hours)	t_4 (4 hours)
1	10.55	4.11	2.00	1.02
2	10.47	4.30	2.15	1.11
3	9.46	3.81	1.78	0.94
4	9.27	3.72	1.92	0.95
5	9.37	3.75	1.95	0.97
6	9.67	4.28	1.96	1.04
7	10.58	3.95	2.30	1.08
8	9.96	3.73	1.86	1.01
9	9.84	3.92	2.00	1.05
10	10.20	4.20	1.96	1.03
11	9.45	4.18	2.18	1.02
12	9.64	4.04	2.08	0.96
13	10.03	4.01	2.08	1.04
14	9.81	3.65	1.97	0.97
15	10.74	4.41	2.07	1.03
16	10.08	3.80	1.86	0.99
17	10.00	3.84	2.07	0.95
18	9.73	3.94	1.93	0.96
19	9.64	4.24	2.11	1.06
20	10.40	4.11	2.07	1.01
21	10.34	4.20	2.21	1.14
22	10.09	4.35	1.91	1.07
23	9.51	3.74	1.87	0.99
24	9.63	3.77	1.96	1.01

Table E
Ramus Height of 20 Boys

Boy	t_1 (age 8)	t_2 (age $8\frac{1}{2}$)	t_3 (age 9)	t_4 (age $9\frac{1}{2}$)
1	47.8	48.8	49.0	49.7
2	46.4	47.3	47.7	48.4
3	46.3	46.8	47.8	48.5
4	45.1	45.3	46.1	47.2
5	47.6	48.5	48.9	49.3
6	52.5	53.2	53.3	53.7
7	51.2	53.0	54.3	54.5
8	49.8	50.0	50.3	52.7
9	48.1	50.8	52.3	54.4
10	45.0	47.0	47.3	48.3
11	51.2	51.4	51.6	51.9
12	48.5	49.2	53.0	55.5
13	52.1	52.8	53.7	55.0
14	48.2	48.9	49.3	49.8
15	49.6	50.4	51.2	51.8
16	50.7	51.7	52.7	53.3
17	47.2	47.7	48.4	49.5
18	53.3	54.6	55.1	55.3
19	46.2	47.5	48.1	48.4
20	46.3	47.6	51.3	51.8

The following points should be noted.

(1) The $y_{i,j}$ data must be in consecutive columns as in Tables C–E, starting with c1 and not extending beyond c10; for example, in Table E there are four columns of data that should be entered in columns c1-c4, respectively, of the worksheet. In particular, c1 contains the numbers 47.8, 46.4, 46.3, 45.1, ..., 46.3.

(2) The number of columns in **(1)** is denoted by k, and it is equal to the number of time-points $t_1, t_2, \ldots, t_k$ at which each item is observed. Note that $k = 4$ in Tables D and E.

(3) The sample size is denoted by m and $m = 24$ in Table D and $m = 20$ in Table E.

(4) The number of unknown parameters in the growth curve model is denoted by p, and $p = 3$ for the model in (7.5.7).

(5) The growth curve model must be a <u>linear model</u> (i.e., linear in the unknown parameters). In Example 7.5.1 it is linear in t, and is given by $\mu_Y(t) = \alpha + \beta t$; in Example 7.5.2 it is quadratic in t, and is given by $\alpha + \beta t + \gamma t^2$.

(6) The $\boldsymbol{X}$ matrix has size k by p where p is the number of unknown parameters in the growth model. The first column of $\boldsymbol{X}$ is a column of 1's if the linear model contains an intercept term α. For a general 3rd degree polynomial model, the $\boldsymbol{X}$ matrix has four columns – the first column has all 1's, the second column has elements t_i, the third column has elements t_i^2, and the fourth column has elements t_i^3.

(7) The columns of the $\boldsymbol{X}$ matrix must be in consecutive columns of the worksheet, starting with `c11` and not extending beyond `c16`.

(8) The vector $\boldsymbol{a}$ in $\boldsymbol{\theta} = \boldsymbol{a}^T\boldsymbol{\beta}$ must be in column c20.

(9) To use the macro, get into Minitab, retrieve the data, make sure that the data are in the proper columns, and finally type `execute 'b:\macro\growth'` .

We illustrate an application of this macro by using Example 7.5.6, where an investigator wants to establish a growth curve for the ramus bone in young boys. A simple random sample of 20 boys was selected and the ramus height for each boy was measured (in millimeters) at ages 8.0, 8.5, 9.0, and 9.5 years. The data are given in Table E above and are also stored in the file **ramus.mtp** on the data disk. We first retrieve this file. We compute a 95% two-sided confidence interval for β, the **average population growth rate** of the ramus bone, where we assume that the **population growth curve** is

$$\mu_Y(t) = \alpha + \beta t$$

Note that, for this example $p = 2$, $m = 20$, and $k = 4$. Also the **X** matrix is

$$\boldsymbol{X} = \begin{bmatrix} 1 & t_1 \\ 1 & t_2 \\ \vdots & \vdots \\ 1 & t_k \end{bmatrix} = \begin{bmatrix} 1 & 8.0 \\ 1 & 8.5 \\ 1 & 9.0 \\ 1 & 9.5 \end{bmatrix} \tag{M7.5.1}$$

The two columns of the matrix $\boldsymbol{X}$ must be entered into columns `c11` and `c12`, respectively, of the worksheet before invoking the macro. Further, note that $\boldsymbol{a}^T = [0 \;\; 1]$, so column c20 must contain 0 as its first element and 1 as its second element. You can use the `set` command to create columns `c11`, `c12`, and c20. Before invoking the macro, be sure that all the relevant data are in the proper columns. The command to start the macro is

COMMAND TO START THE GROWTH CURVE MACRO

```
MTB > execute 'b:\macro\growth'
```

Minitab responds with:

```
-------------------------------------------------------------------------

  This program will calculate a 1-alpha confidence interval for
  a linear combination of the population parameters, i.e., for theta
  where

        theta = a(0) alpha + a(1) beta + ...  .

  The quantities a(0), a(1), ... are the elements of the vector  a
  which must be in column c20.

  The y(i,j) data must be organized as in Tables C, D, and E, in
  consecutive columns c1-ck and k must be less than or equal to 10.

  Let m denote the number of rows of y(i,j); i.e., the sample size.
  m must be less than or equal to 60.

  The X matrix has k rows and p columns where p is the number
  of unknown parameters in the vector beta. The p columns of the
  matrix X must be in consecutive columns c11, ..., c(10+p), where p
  must be less than or equal to 6; i.e., if X is a matrix with 4
  rows (k=4) and 2 columns as in Equation (7.5.41), then p=2, and
  the columns of the X matrix are in c11 and c12.

*************************************************************************

  Input the value of m, the sample size.
DATA> 20

  Input the value of k, the number of distinct values of t;  k is
  also the number of rows in the matrix X.
DATA> 4
```

```
  Input the value of p, the number of elements in the beta vector;
  p is also the number of columns in the matrix X.
DATA> 2

  Input the value of the confidence coefficient 1-alpha.
DATA> .95

====================================================================

  The estimated regression coefficients, in order, are


                    33.74750
                    1.866023


  The estimated value of theta is      1.866023
  and its standard error is           0.2605990


  For a two-sided confidence interval for theta with
  confidence coefficient equal to  0.950

  the lower confidence bound is      1.320580   and
  the upper confidence bound is      2.411465

--------------------------------------------------------------------
```

From this we get $\hat{\beta} = 1.87$ millimeters, and the 95% confidence statement for β is

$$C[1.32 \le \beta \le 2.41] = 0.95$$

Problems

M7.5.1 Work Problem 7.5.1 using the macro described in this section.

M7.5.2 Work Exercise 7.6.2 using the macro described in this section.

Chapter 8

Alternate Assumptions for Regression

8.1 Overview

No computing instructions are needed in this section.

8.2 Straight Line Regression with Unequal Subpopulation Standard Deviations

In this section we demonstrate how Minitab can be used to perform a weighted regression analysis for a straight line model. We illustrate by using the data of Example 8.2.1, which are given in Table 8.2.1 and are also stored in the file **carbmon.mtp** on the data disk. The response variable Y is named `CO` and it is in column `c1`, and the predictor variable X is named `Cars` and it is in column `c2`.

It is known that the subpopulation standard deviations $\sigma_Y(x)$ are not all the same, but the investigator expects the weighted regression assumptions in Box 8.2.1 to hold with $\sigma_Y(x) = \sigma_0 g(x)$, where σ_0 is an unknown constant and $g(x) = \sqrt{x}$. We illustrate the Minitab commands for estimating β_0 and β_1, and for computing standard errors of the estimates, using a weighted least squares regression analysis where the user supplies the weights. The weights in Example 8.2.1 are $1/[g(x)]^2 = 1/x$. We compute the weights, put them in column c3, and name this column. Then we print columns c1, c2, and c3, and check the numbers. The commands are

```
let c3=1/c2
name c3 'weights'
print c1 c2 c3
```

Minitab responds with:

```
--------------------------------------------------------------------------

 ROW     CO   Cars    weights

   1   5817    873   0.0011455
   2   1063    109   0.0091743
   3   2616    398   0.0025126
   4   2018    353   0.0028329
   5   3147    506   0.0019763
   6   7210   1026   0.0009747
   7   4339    862   0.0011601
   8   5153    742   0.0013477
   9   4450    786   0.0012723
  10   5591    896   0.0011161
  11   2747    377   0.0026525
  12   3712    720   0.0013889
  13   2354    655   0.0015267

--------------------------------------------------------------------------
```

Next we give the command and subcommands to perform a weighted regression of Y on X using the weights in c3. Note that, to compute a weighted regression analysis, we use the usual regression command `regress` with the subcommand `weights c3`, which tells Minitab that a weighted regression is to be computed using the weights in column c3. If you also want to compute the matrix $C^{(w)}$, you can do so with the subcommand `xpxi m1`, and this puts $C^{(w)}$ into `m1`. Thus the command and subcommands for a weighted regression of the carbon monoxide data are

COMMAND AND SUBCOMMANDS FOR WEIGHTED REGRESSION

```
MTB > regress c1 1 c2;
SUBC> weights c3;
SUBC> xpxi m1.
```

Minitab responds with:

```
------------------------------------------------------------------------

The regression equation is

CO = 372 + 5.47 Cars

Predictor        Coef        Stdev    t-ratio        p
Constant        371.6        297.6       1.25    0.238
Cars           5.4662       0.5569       9.82    0.000

Analysis of Variance

SOURCE        DF         SS          MS         F        p
Regression     1      74445       74445     96.36    0.000
Error         11       8499         773
Total         12      82944
------------------------------------------------------------------------
```

We print the matrix $C^{(w)}$ which is stored in m1, using the command

```
print m1
```

Minitab responds with:

```
------------------------------------------------------------------------

 MATRIX M1

 114.5957   -0.1794
  -0.1794    0.0004
------------------------------------------------------------------------
```

Note that the output has the same form as the output from an ordinary regression analysis except `s`, `R-sq`, and `R-sq(adj)` are not computed.

Problems

M8.2.1 Work Problems 8.2.1 through 8.2.4 using the commands discussed in this section.

M8.2.2 Work Exercise 8.4.1 using Minitab to do the computations.

8.3 Straight Line Regression–Theil's Method

In this section we explain the Minitab macro named **theil** that can be used to obtain point estimates and confidence intervals for the linear combination

$$\theta = a_0\beta_0 + a_1\beta_1$$

in the straight line regression model

$$\mu_Y(x) = \beta_0 + \beta_1 x$$

when assumptions in Box 8.3.1 are satisfied. The Minitab commands for this macro are stored in the files **theil.mtb**, **theil1.mtb**, and **theil2.mtb** in the subdirectory `macro` on the data disk. We illustrate this macro by using the data of Example 8.3.1, which are given in Table 8.3.3 and are also stored in the file **profsal.mtp** on the data disk. In this example the response variable Y is `salary` and the predictor variable X is `yrsexp`. They are stored in columns `c1` and `c2`, respectively.

We compute a point estimate for $\mu_Y(10)$ and a confidence interval for it with confidence coefficient as close to 90% as possible. To use the macro **theil**, you must have the Y values in column `c1` and the X values in column `c2`. You will be asked to input the coefficients a_0 and a_1 which, for this example, are $a_0 = 1$ and $a_1 = 10$. You will also be asked to input the desired confidence coefficient for a two-sided confidence interval for $\theta = a_0\beta_0 + a_1\beta_1$. Exact confidence intervals are available by Theil's method for only a selected set of confidence coefficients, so the macro will automatically choose an allowable confidence coefficient that is closest to the desired confidence coefficient.

Retrieve the file **profsal.mtp** and start the macro with the following command.

MACRO COMMAND FOR THEIL'S METHOD

```
MTB > execute 'b:\macro\theil'
```

Minitab responds with:

```
------------------------------------------------------------------------

  This program computes a point estimate and a confidence interval
  for

      theta = a(0) beta(0) + a(1) beta(1)

  using Theil's method. Assumptions in Box 8.3.1 are presumed
  valid.

  The Y data must be in column c1.

  The X data must be in column c2.

**********************************************************************

  Input a(0) and a(1), the coefficients of beta(0) and of beta(1),
  in this order.
DATA> 1 10

  Input the value of 1-alpha, the desired confidence coefficient
  for a two-sided confidence interval for theta.
DATA> .90

======================================================================

  The point estimate of theta is     46.44444


  For a two-sided confidence interval for theta with confidence
  coefficient equal to  0.875 (this is the value that is closest
  to the desired value of  0.900)

  the lower confidence bound is     44.00000  and
  the upper confidence bound is     48.33333

------------------------------------------------------------------------
```

Thus the point estimate of $\mu_Y(10)$ is 46.4444, and a two-sided confidence interval with confidence coefficient as close to 0.90 as possible is given by the confidence statement

$$C[44.0000 \leq \mu_Y(10) \leq 48.3333] = 0.875$$

Problems

M8.3.1 In Example 8.3.1 find a point estimate and a confidence interval for β_0 with confidence coefficient as close to 90% as possible using assumptions in Box 8.3.1.

M8.3.2 In Example 8.3.1 find a point estimate and a confidence interval for β_1 with confidence coefficient as close to 90% as possible using assumptions in Box 8.3.1.

M8.3.3 Work Problem 8.3.1 by using the macro explained in this section.

Chapter 9

Nonlinear Regression

9.1 Overview

No computing instructions are needed in this section.

9.2 Some Commonly Used Families of Nonlinear Regression Functions

No computing instructions are needed in this section.

9.3 Statistical Assumptions and Inferences for Nonlinear Regression

In this section we describe how to use a Minitab macro for nonlinear regression provided by us on the data disk. The name of this macro is **nonlin**, and the Minitab instructions for this macro are stored in the files **nonlin.mtb**, **nonlin1.mtb**, **nonlin2.mtb**, **nonlin3.mtb**, **nonlin4.mtb**, and **nonline.mtb** in the subdirectory macro on the data disk. This macro may be used to fit any of the four nonlinear models listed below in

Table F. Each one of these models involves only a *single* predictor variable X.

Table F
Nonlinar Regression Models
Included in the Minitab Macro nonlin

$$(1)\quad \mu_Y(x) = \frac{1}{\beta_1 + \beta_2 x + \beta_3 x^2}$$

$$(2)\quad \mu_Y(x) = \beta_1[1 - e^{-e^{-(\beta_2+\beta_3 x)}}]$$

$$(3)\quad \mu_Y(x) = \frac{1}{1 + e^{-(\beta_1+\beta_2 x)}}$$

$$(4)\quad \mu_Y(x) = \beta_1 + \beta_2 e^{-\beta_3 x}$$

Before proceeding further we **caution you that this macro has been provided for illustrative purposes only and that it is not a general purpose program for fitting nonlinear regression functions. It cannot be used to fit models other than those in Table F. It has been tested only on the examples and problems discussed in this chapter. For other problems or other models, you should use nonlinear fitting programs available in some of the other statistical software packages such as SAS, BMDP, SPSS, etc.**

Before invoking the nonlinear regression macro, you must retrieve the data from the appropriate data file and check them. During execution of the macro you will be asked to input the following information:

(1) The number of the column where the values of Y are stored.

(2) The number of the column where the values of X are stored.

(3) The number of β parameters in the model.

(4) Initial guesses for the parameter values (which will be used as starting values for the iterations).

(5) The number of the column where these initial β values can be stored.

(6) The number of iterations (we recommend at least 20 as a rule of thumb; some problems will need more iterations).

(7) The confidence coefficient $1-\alpha$ for computing confidence intervals.

(8) The number of the nonlinear model in Table F that is being used (the macro will list these models for your convenience).

When all required input values have been entered, the program will automatically perform the requested number of iterations. Summary results are displayed on the screen after each iteration. The final results (assuming that the program has converged to the right estimates) are displayed on the screen. These include a table of parameter estimates, their (approximate) standard errors, and lower and upper bounds corresponding to one-at-a-time two-sided $1-\alpha$ confidence intervals. The macro also lists the observed Y values and the corresponding fitted values. A plot of the observed Y values against the X values is given, and superimposed on this is also a plot of the fitted Y values against the X values. This plot can be used to judge the adequacy of the fit visually.

To illustrate the macro, we use the data from Example 9.3.1 which are given in Table 9.3.1 and are also stored in the file **light.mtp** on the data disk. You should retrieve these data and check them before proceeding with the analysis. We will fit model (4) in Table F to these data. As initial guesses we use $\beta_1 = 0.0$, $\beta_2 = 2.0$, and $\beta_3 = 0.5$. We ask Minitab to store these initial guesses in column c3. For the number of iterations, we choose 10. Also we use 0.95 for the confidence coefficient for one-at-a-time two-sided confidence intervals for each β parameter. The command to start the macro is

COMMAND FOR STARTING THE MACRO NONLIN

```
MTB > execute 'b:\macro\nonlin'
```

Minitab responds with:

```
--------------------------------------------------------------------------------

       ** Nonlinear Regression Macro - Iterative Solutions **

  Acknowledgement:

  Adapted from a macro written by
               Gary Stork
               Quality Assurance Specialist
               Ford Motor Company
               17101 Rotunda Drive - Box 1578-A
               Dearborn, Mi 48121

     ** statlib macro NONLIN.MTB Wed Aug 26 03:01:58 EDT 1992 **
  Name: NONLIN.MTB
**********************************************************************

  Enter the number of the column containing the Y values.
DATA> 1

  Enter the number of the column containing the X values.
DATA> 2

  Enter the number of beta parameters.
DATA> 3

  Enter the number of the column for storing the initial values.
DATA> 3

  Enter the initial values for beta1, beta2, ..., etc., in order.
DATA> 0 2 .5

  Enter the number of iterations to be performed.
DATA> 10

  Enter the confidence coefficient for one-at-a-time two-sided
  confidence intervals.
DATA> .95
```

```
 +++++++++++++++++++++++++++++++++++++++++++++++++++++++++++++

 We have provided 4 nonlinear regression models in this macro.
 They are:

 Model 1: y = 1/(beta1 + beta2*x + beta3*x^2)

 Model 2: y = beta1*{1-exp[-exp(-(beta2+beta3*x))]}

 Model 3: y = 1/[1+exp(-(beta1+beta2*x))]

 Model 4: y = beta1+beta2*exp(-beta3*x)

 +++++++++++++++++++++++++++++++++++++++++++++++++++++++++++++

 Enter the model number ( 1 through 4 )
DATA> 4

======================================================================

 Iteration number =      0
    estimate of beta1 =             0
    estimate of beta2 =      2.000000
    estimate of beta3 =     0.5000000

 SSE =      1.688465

Ready to continue ? (y = yes)
y

 Iteration number =      1
    estimate of beta1 =     0.1197266
    estimate of beta2 =      2.631557
    estimate of beta3 =     0.7511375

 SSE =    0.4774066

Ready to continue ? (y = yes)
y

 Iteration number =      2
    estimate of beta1 =    0.03221978
    estimate of beta2 =      2.718837
    estimate of beta3 =     0.6772598

 SSE =    0.4611079

Ready to continue ? (y = yes)
y
```

```
  Iteration number =      3
     estimate of beta1 =   0.02894145
     estimate of beta2 =     2.723149
     estimate of beta3 =    0.6829071

  SSE =     0.4604272

 Ready to continue ? (y = yes)
y

  Iteration number =      4
     estimate of beta1 =   0.02874978
     estimate of beta2 =     2.723279
     estimate of beta3 =    0.6827589

  SSE =     0.4604271

 Ready to continue ? (y = yes)
y

  Iteration number =      5
     estimate of beta1 =   0.02875370
     estimate of beta2 =     2.723276
     estimate of beta3 =    0.6827627

  SSE =     0.4604271

 Ready to continue ? (y = yes)
y

  Iteration number =      6
     estimate of beta1 =   0.02875357
     estimate of beta2 =     2.723276
     estimate of beta3 =    0.6827626

  SSE =     0.4604271

 Ready to continue ? (y = yes)
y

  Iteration number =      7
     estimate of beta1 =   0.02875350
     estimate of beta2 =     2.723277
     estimate of beta3 =    0.6827625

  SSE =     0.4604271

 Ready to continue ? (y = yes)
y
```

```
   Iteration number =      8
      estimate of beta1 =   0.02875359
      estimate of beta2 =     2.723276
      estimate of beta3 =    0.6827626

   SSE =    0.4604271

 Ready to continue ? (y = yes)
y

   Iteration number =      9
      estimate of beta1 =   0.02875358
      estimate of beta2 =     2.723277
      estimate of beta3 =    0.6827626

   SSE =    0.4604271

 Ready to continue ? (y = yes)
y


   Parameter estimates and approximate standard errors (ASE)
   after the final iteration are:

         parameter          estimate                ASE

             beta1        0.02875358          0.1715195
             beta2          2.723276          0.2105402
             beta3         0.6827626          0.1416589

   For two-sided one-at-a-time confidence intervals for the beta
   parameters with confidence coefficient approximately equal to
   0.950 the lower and the upper bounds are given below.

         parameter       lower bound        upper bound

             beta1        -0.3592508          0.4167580
             beta2          2.247001           3.199552
             beta3         0.3623078           1.003217

   The estimated subpopulation standard deviation is    0.2261826


 Ready to continue ? (y = yes)
y
```

The observed values (Y) and the fitted values (Yhat) are

Observed Y	Fitted Yhat
2.860000	2.752030
2.640000	2.752030
1.570000	1.404606
1.240000	1.404606
0.4500000	0.7238606
1.020000	0.7238606
0.6500000	0.3799350
0.1800000	0.3799350
0.1500000	0.2061772
0.01000000	0.2061772
0.04000000	0.1183915
0.3600000	0.1183915

Ready to continue ? (y = yes)
y

A plot of the observed Y values against the X values and a superimposed plot of the fitted Yhat values against the X values is given below.

```
        -
     3.0+
        - 3
        - A
        -
        -
     2.0+
        -
        -          A
        -          2
        -          A
     1.0+                    A
        -                    2
        -                              A
        -                    A         2                   A
        -                              A         3         2
     0.0+                                        A         A
          +---------+---------+---------+---------+---------+------
         0.0       1.0       2.0       3.0       4.0       5.0
       A = Y vs. X                B = Yhat vs. X
-------------------------------------------------------------------------------
```

Compare these results with the results given in Exhibit 9.3.1 (which were obtained using SAS). The slight differences in the results are due to different algorithms used for the computations, which lead to different degrees of rounding errors.

The plot of observed and fitted values against X was done using the Minitab command `mplot`. Points labeled `A` form the plot of Y against X, whereas points labeled `B` form the plot of $\hat{Y}$ against X. If two or more points overlap (usually due to the low resolution of the plot), then Minitab uses the number of overlapping points as the 'plot symbol'. In the plot above, none of the `B` symbols are visible because they overlap with some of the `A` points, and so Minitab has used either a 2 or a 3 as the plot symbol as appropriate. In particular, we see that the fit is quite good in this problem.

Notice that the macro prints the estimates of the β parameters and the SSE at the end of each iteration. This allows you to monitor the progress of the iteration procedure and decide whether or not convergence has been attained in the number of iterations specified. In the computer output above we see that convergence has been attained by the end of the third iteration. This is partly because the choice of the initial values for this problem turned out to be good. This will not always be the case, and most problems require many more iterations. If it appears that convergence has not been attained at the end of the specified number of iterations, then you should rerun the macro using the parameter values from the last iteration as your initial guess.

After printing the results at the end of an iteration, the macro pauses to allow you time to study the numbers. It asks you whether you are ready for the macro to continue with the next iteration. You type `y` for **yes** when you are ready for the macro to continue.

There is no guarantee that the result given by the program is in fact correct even when convergence appears to have been attained. This is true of most nonlinear regression programs. You must check the results carefully. It is often useful to try different starting values and see if the final results are the same. If you encounter problems, we recommend that you consult a statistician.

Problems

M9.3.1 Consider the experiment discussed in Problem 9.3.3, where we provided a SAS output containing the results of a nonlinear regression analysis for that problem. Now solve Problem 9.3.3 using the Minitab macro **nonlin**. For starting values for β_1, β_2, and β_3, you may use the values 0.8, -0.67, and 0.16, respectively. You may use other starting values, but convergence cannot be guaranteed if starting values are chosen arbitrarily.

M9.3.2 Rework Problem 9.3.2 using the Minitab macro **nonlin**. Use $\beta_1 = 2, \beta_2 = -1$, and $\beta_3 = 14$ as starting values.

9.4 Linearizable Models

All Minitab commands needed in this section have already been discussed.

Problems

M9.4.1 Use the macro **nonlin** to fit the nonlinear model in Example 9.4.1. Use the estimates obtained from the linearization approach as the initial guesses for β_1 and β_2.

M9.4.2 Rework Example 9.4.1 using the Minitab macro **nonlin** with $\beta_1 = -3.0$ and $\beta_2 = 150.0$ as the initial guesses.

M9.4.3 Use the macro **nonlin** to fit the nonlinear model in Problem 9.4.1. Use the estimates obtained from the linearization approach as the initial guesses for β_1, β_2, and β_3.

Answers and Solutions

M1.1.1 The Minitab commands are

```
MTB > set c1
DATA> 1.5 1.9 1.2 2.1 1.6 1.7
MTB > set c2
DATA> 600 590 710 560 610 700
MTB > set c3
DATA> 34.5 43.9 30.3 31.7 42.1 39.4
DATA> end
MTB >
```

M1.1.2 The Minitab commands are

```
MTB > name c1 'Y'
MTB > name c2 'X'
MTB > name c3 'Z'
```

M1.1.3 The Minitab command and response are

```
MTB > print c1 c2 c3
```

```
 ROW      Y      X      Z

   1    1.5    600   34.5
   2    1.9    590   43.9
   3    1.2    710   30.3
   4    2.1    560   31.7
   5    1.6    610   42.1
   6    1.7    700   39.4
```

M1.1.4 To write the data to the ASCII file named **prob111.dat** and store it in the working directory on the `C` drive, the command is

```
MTB > write 'c:prob111.dat' c1 c2 c3
```

M1.1.5 To save the contents of the entire worksheet in a portable Minitab file named **prob111.mtp** and store in the working directory on the `C` drive, the command is

```
MTB > save 'c:prob111';
SUBC> portable.
```

M1.1.6 The Minitab command and response are

```
MTB > describe c1 c2 c3

               N     MEAN   MEDIAN   TRMEAN    STDEV   SEMEAN
Y              6    1.667    1.650    1.667    0.314    0.128
X              6    628.3    605.0    628.3     61.8     25.2
Z              6    36.98    36.95    36.98     5.63     2.30

             MIN      MAX       Q1       Q3
Y          1.200    2.100    1.425    1.950
X          560.0    710.0    582.5    702.5
Z          30.30    43.90    31.35    42.55
```

Thus we get $\hat{\mu}_Y = 1.667$, $\hat{\sigma}_X = 61.8$, the maximum value of $Z = 43.90$, and the minimum value of $Z = 30.30$.

M1.1.7 The Minitab commands and responses are

```
MTB > retrieve 'b:\table161';
SUBC> portable.

Worksheet retrieved from file: b:\table161.MTP

MTB > info

COLUMN     NAME       COUNT
C1         mpg           10

CONSTANTS USED: NONE
```

M1.1.8 The Minitab command and response are

```
MTB > print c1

mpg
   25.72   25.24   25.19   25.88   26.42   24.48   25.11   24.29   25.06
   25.63
```

M1.1.9 The Minitab command is

```
MTB > name c1 'MPgal'
```

M1.1.10 The Minitab command and response are

```
MTB > describe c1

                 N     MEAN   MEDIAN   TRMEAN    STDEV   SEMEAN
MPgal           10   25.302   25.215   25.289    0.639    0.202

               MIN      MAX       Q1       Q3
MPgal       24.290   26.420   24.915   25.760
```

Thus the mean of MPgal is 25.302 and the standard deviation is 0.639.

M1.1.11 From the Minitab output in **M1.1.10** we find that the minimum value of MPgal is 24.290 and the maximum value is 26.420.

M1.1.12 The Minitab command and response are

```
MTB > retrieve 'b:\table164';
SUBC> portable.

Worksheet retrieved from file: b:\table164.MTP
```

M1.1.13 The Minitab command and response are

```
MTB > info

COLUMN     NAME        COUNT
C1                        30

CONSTANTS USED: NONE
```

M1.1.14 The Minitab command and response are

```
MTB > print c1

C1
    3.48    7.68   12.96    0.65   12.44    0.16    4.34    3.71    4.67
    3.47    5.91    6.73    3.64   10.90    6.37    4.62    9.18   11.67
    8.29    5.19    4.30   11.67    8.63    7.84   11.92   11.16    6.13
   10.21    8.16    3.59
```

M1.1.15 The Minitab command and response are

```
MTB > describe c1

                 N      MEAN    MEDIAN    TRMEAN     STDEV    SEMEAN
C1              30     6.989     6.550     7.056     3.544     0.647

               MIN       MAX        Q1        Q3
C1           0.160    12.960     4.153    10.382
```

So the mean is $\hat{\mu}_Y = 6.989$, the standard deviation is $\hat{\sigma}_Y = 3.544$, and the median is 6.550.

M1.6.1 The Minitab commands and responses are

```
MTB > retrieve 'b:\table164';
SUBC> portable.

Worksheet retrieved from file: b:\table164.MTP
MTB > name c1 'Y'
MTB > print c1

Y
   3.48    7.68   12.96    0.65   12.44    0.16    4.34    3.71    4.67
   3.47    5.91    6.73    3.64   10.90    6.37    4.62    9.18   11.67
   8.29    5.19    4.30   11.67    8.63    7.84   11.92   11.16    6.13
  10.21    8.16    3.59
```

M1.6.2 The Minitab command and response are

```
MTB > describe c1

                 N      MEAN    MEDIAN    TRMEAN     STDEV    SEMEAN
Y               30     6.989     6.550     7.056     3.544     0.647

               MIN       MAX        Q1        Q3
Y            0.160    12.960     4.153    10.382
```

Thus we get $\hat{\mu}_Y = 6.989$.

M1.6.3 We use the following command to obtain a 80% two-sided confidence interval for μ_Y and from this we can obtain a 90% upper confidence bound.

```
MTB > tinterval .80 c1

              N      MEAN     STDEV   SE MEAN    80.0 PERCENT C.I.
C1           30     6.989     3.544     0.647  (   6.140,    7.838)
```

Thus we get the confidence statement $C[\mu_Y \leq 7.838] = 0.90$.

M1.6.4 The Minitab command and response are

```
MTB > ttest 4.5 c1

TEST OF MU = 4.500 VS MU N.E. 4.500

              N      MEAN     STDEV   SE MEAN          T    P VALUE
Y            30     6.989     3.544     0.647       3.85     0.0006
```

Thus the value of t_C is 3.85.

M1.6.5 The P-value is 0.0006.

M1.6.6 The Minitab command and response are

```
MTB > ttest 5.0 c1;
SUBC> alternative = +1.

TEST OF MU = 5.000 VS MU G.T. 5.000

              N      MEAN     STDEV   SE MEAN          T    P VALUE
Y            30     6.989     3.544     0.647       3.07     0.0023
```

So $t_C = 3.07$ and the P-value is 0.0023.

M1.6.7 The Minitab commands and responses are

```
MTB > let k1=sum(c1)
MTB > print k1
K1        209.670
```

M1.6.8 The Minitab commands and responses are

```
MTB > let k2=ssq(c1)
MTB > print k2
K2        1829.58
```

M1.6.9 The Minitab commands and responses are

```
MTB > let k11=ssq(c1-mean(c1))
MTB > print k11
K11       364.193
```

M1.6.10 The Minitab commands and responses are

```
MTB > let k10=c1(1)/c1(13)+c1(4)*c1(6)
MTB > print k10
K10       1.06004
```

M1.8.1 Answers to parts **(a)**–**(m)** are given below.

(a) We read the 7×3 matrix X into `m11`.

```
MTB > read 7 3 m11
DATA> 12 28 21
DATA> 14 31 46
DATA> 20 21 31
DATA> 11 19 21
DATA> 16 13 34
DATA> 39 26 30
DATA> 25 37 15
      7 ROWS READ
```

Next we read the 7×1 matrix $\boldsymbol{y}$ into `m12`.

```
MTB > read 7 1 m12
DATA> 9
DATA> 13
DATA> 28
DATA> 6
DATA> 32
DATA> 16
DATA> 24
      7 ROWS READ

MTB > print m11 m12
 MATRIX M11

  12   28   21
  14   31   46
  20   21   31
  11   19   21
  16   13   34
  39   26   30
  25   37   15

 MATRIX M12

   9
  13
  28
   6
  32
  16
  24
```

(b) We transpose X and put it into `m1`, then we compute $X^T X$ and put it into `m2`. Finally, we print `m1` and `m2`.

```
MTB > transpose m11 m1
MTB > multiply m1 m11 m2
MTB > print m1 m2
```

```
 MATRIX M1

  12   14   20   11   16   39   25
  28   31   21   19   13   26   37
  21   46   31   21   34   30   15


 MATRIX M2

  3263   3546   3836
  3546   4761   4841
  3836   4841   6240
```

(c) We compute $X^T y$, put the result into m3, and print it.

```
MTB > multiply m1 m12 m3
MTB > print m3

 MATRIX M3

  2652
  3077
  3709
```

(d) We compute $(X^T X)^{-1}$, put the result into m4, and print m4.

```
MTB > invert m2 m4
MTB > print m4

 MATRIX M4

 0.00171927 -0.00097482 -0.00030065
-0.00097482  0.00154740 -0.00060121
-0.00030065 -0.00060121  0.00081150
```

(e) We compute $(XX)^{-1} X^T y$, put the result into m5, and print it.

```
MTB > multiply m4 m3 m5
MTB > print m5

 MATRIX M5

 0.444897
-0.053762
 0.362602
```

(f) We transpose y (i.e., m12) and put the result into m6. Then we calculate $y^T y$ (i.e., m6 times m12) and put the result into m7. The result is a 1×1 matrix, i.e., a single number (sometimes called a *scalar*), and Minitab prints it automatically.

```
MTB > transpose m12 m6
MTB > multiply m6 m12 m7
ANSWER =        2926.0000
```

(g) Note that $y^T[I - X(X^T X)^{-1}X^T]y = y^T y - y^T X(X^T X)^{-1}X^T y$. We compute $y^T X$ by transposing m3. The result is put into m8.

```
MTB > tranpose m3 m8
```

Then we multiply $y^T X$ in m8 and $(X^T X)^{-1}X^T y$ in m5 to get $y^T X(X^T X)^{-1}X^T y$. The result is stored in m9. Since the result is a 1×1 matrix, i.e., a single number, Minitab prints it automatically.

```
MTB > multiply m8 m5 m9
ANSWER =        2359.3320
```

Finally we subtract m9 from m7 to get $y^T y - y^T X(X^T X)^{-1}X^T y$ using the command

```
MTB > subt m9 m7 m10
MTB > print m10
ANSWER =         566.6680
```

The result is stored in m10. Since the result is a scalar, Minitab prints it automatically.

(h) First we construct a column, say c7, with seven elements, each equal to one, with the command

```
MTB > set c7
DATA> 1 1 1 1 1 1 1
DATA> end
```

Next we create the 7×7 identity matrix $\boldsymbol{I}$, denoted by m7, by constructing a diagonal matrix with c7 as the diagonal. We then print m7. The commands and the Minitab response are

```
MTB > diagonal is c7 put into m7
MTB > print m7
 MATRIX M7

   1   0   0   0   0   0   0
   0   1   0   0   0   0   0
   0   0   1   0   0   0   0
   0   0   0   1   0   0   0
   0   0   0   0   1   0   0
   0   0   0   0   0   1   0
   0   0   0   0   0   0   1
```

We obtain $(1/7)\boldsymbol{J}$ with the commands

```
MTB > let k11=1/7
MTB > define k11 7 by 7 matrix m8
```

where k11 $= 1/7 = 0.142857$. From these results you can obtain $\boldsymbol{E}$ using the following commands. $\boldsymbol{E}$ is stored in m9.

```
MTB > subt m8 m7 m9
MTB > print m9
```

```
 MATRIX M9

 0.857143 -0.142857 -0.142857 -0.142857 -0.142857 -0.142857 -0.142857
-0.142857  0.857143 -0.142857 -0.142857 -0.142857 -0.142857 -0.142857
-0.142857 -0.142857  0.857143 -0.142857 -0.142857 -0.142857 -0.142857
-0.142857 -0.142857 -0.142857  0.857143 -0.142857 -0.142857 -0.142857
-0.142857 -0.142857 -0.142857 -0.142857  0.857143 -0.142857 -0.142857
-0.142857 -0.142857 -0.142857 -0.142857 -0.142857  0.857143 -0.142857
-0.142857 -0.142857 -0.142857 -0.142857 -0.142857 -0.142857  0.857143
```

(i) We multiply $\boldsymbol{E}$ by $\boldsymbol{E}$, print the result, and confirm that $\boldsymbol{EE} = \boldsymbol{E}$. The commands are

```
MTB > multiply m9 m9 m10
MTB > print m10
 MATRIX M10

 0.857143 -0.142857 -0.142857 -0.142857 -0.142857 -0.142857 -0.142857
-0.142857  0.857143 -0.142857 -0.142857 -0.142857 -0.142857 -0.142857
-0.142857 -0.142857  0.857143 -0.142857 -0.142857 -0.142857 -0.142857
-0.142857 -0.142857 -0.142857  0.857143 -0.142857 -0.142857 -0.142857
-0.142857 -0.142857 -0.142857 -0.142857  0.857143 -0.142857 -0.142857
-0.142857 -0.142857 -0.142857 -0.142857 -0.142857  0.857143 -0.142857
-0.142857 -0.142857 -0.142857 -0.142857 -0.142857 -0.142857  0.857143
```

Notice that m10 is the same as m9.

(j) We calculate $\boldsymbol{y}^T[\frac{1}{7}\boldsymbol{J}]\boldsymbol{y}$ using the Minitab commands below.

```
MTB > multiply m6 m8 m13
MTB > multiply m13 m12 m14
ANSWER =       2340.5715
```

We then calculate $7\bar{y}^2$ and obtain 2340.57 as the result. The commands are

```
MTB > copy m12 c99
MTB > let k12=7*mean(c99)*mean(c99)
MTB > print k12
K12       2340.57
```

This verifies the equation $y^T[\frac{1}{7}J]y = 7\bar{y}^2$ (to within rounding error).

(k) We calculate $\frac{1}{7}\mathbf{1}^T y$ using the following commands.

```
MTB > set c98
DATA> 1 1 1 1 1 1 1
MTB > transpose c98 m13
MTB > multiply m13 m12 m14
ANSWER =         128.0000
MTB > let k11=1/7
MTB > multiply k11 m14 m15
ANSWER =          18.2857
```

The result is 18.2857. Next we calculate $\bar{y}$ using the following command

```
MTB > let k1=mean(c99)
MTB > print k1
K1        18.2857
```

Notice that we obtained the same result in both cases.

(l) We calculate $y^T Ey$ using the following commands.

```
MTB > multiply m9 m12 m13
MTB > multiply m6 m13 m14
ANSWER =         585.4285
```

The result is 585.4285. Next we calculate SSY and print it. The commands are

```
MTB > let k14=ssq(c99-mean(c99))
MTB > print k14
K14       585.429
```

This verifies the equality $y^T Ey = SSY$ (to within rounding error).

(m) First create the matrix J and store it in m10 using the following command.

```
MTB > define 1 in 7 by 7 matrix m10
```

You can view this matrix using the command **print m10** . Now multiply $\boldsymbol{E}$ in **m9** by J in m10, store the result in **m14**, and print it. The commands are

```
MTB > multiply m9 m10 m14
MTB > print m14
```

```
MATRIX M14

-0.000000030 -0.000000030 -0.000000030 -0.000000030 -0.000000030 -0.000000030 -0.000000030
-0.000000030 -0.000000030 -0.000000030 -0.000000030 -0.000000030 -0.000000030 -0.000000030
-0.000000030 -0.000000030 -0.000000030 -0.000000030 -0.000000030 -0.000000030 -0.000000030
-0.000000030 -0.000000030 -0.000000030 -0.000000030 -0.000000030 -0.000000030 -0.000000030
-0.000000030 -0.000000030 -0.000000030 -0.000000030 -0.000000030 -0.000000030 -0.000000030
-0.000000030 -0.000000030 -0.000000030 -0.000000030 -0.000000030 -0.000000030 -0.000000030
-0.000000030 -0.000000030 -0.000000030 -0.000000030 -0.000000030 -0.000000030 -0.000000030
```

We notice that the result is a matrix of zeros (except for rounding error in the eighth decimal place) which verifies the equation $\boldsymbol{EJ} = 0$.

M1.9.1 The Minitab commands and response are

```
MTB > retrieve 'b:\bivgauss';
SUBC> portable.
Worksheet retrieved from file: b:\bivgauss.MTP
```

M1.9.2 The Minitab command and response are

```
MTB > info

COLUMN     NAME     COUNT
C1         X1        1000
C2         X2        1000

CONSTANTS USED: NONE
```

From the `info` command we see that the file **bivgauss.mtp** contains two columns of data with 1000 observations in each column. Column `c1` has the name $X1$ and column c2 has the name $X2$.

M1.9.3 The Minitab command and response are

```
MTB > histogram c1


Histogram of X1   N = 1000
Each * represents 5 obs.

Midpoint   Count
     -10       3  *
      -8       8  **
      -6      19  ****
      -4      65  *************
      -2     118  ************************
       0     163  *********************************
       2     224  *********************************************
       4     177  ************************************
       6     128  **************************
       8      56  ************
      10      31  *******
      12       6  **
      14       2  *
```

M1.9.4 The Minitab command and response are

```
MTB > histogram c2


Histogram of X2   N = 1000
Each * represents 5 obs.

Midpoint   Count
     -25       2  *
     -20      14  ***
     -15      40  ********
     -10      77  ****************
      -5     157  ********************************
       0     186  **************************************
       5     201  *****************************************
      10     167  **********************************
      15      92  *******************
      20      41  *********
      25      20  ****
      30       2  *
      35       1  *
```

M1.9.5 The Minitab command and response are

```
MTB > plot c1 c2


        -
X1      -
        -                   *  *
        -                    2** * *
      10+             2 **    2*  222 *32322*  * *   **
        -           *      **242*5*3*242424*432  2      2 *    *
        -         *   3 **233974446978733+496*24*2**   2
        -        **** 4*2335459556+668+859948 5846*3224
        -     2     *2*23427768796+779++++5+6+756343*4**22
       0+        *  *  *22465372+879745+5955+8254 * 53  *2
        -        * 2***4**2* 56434646767785343*253*   **   *
        -           *  ** *222423*32436442 34 3 2 *2*     *
        -              *  *   *  * 2 *22*  2** **    *
        -              *                ****  *     2
     -10+                    2          *
        -
        -
         ------+---------+---------+---------+---------+---------+X2
              -24       -12        0         12        24        36
```

M1.9.6 The Minitab command and response are

```
MTB > plot c2 c1


        -
X2      -                                        *
        -                      *    * *          *
        -                 *    *2 ***2 3*2* **  ** * *
      20+         * * *  2 2   *4*2* 2*4*43**   *    *
        -           * *  **  24234*2*454536*6 *3***     **
        -       *    *  2**43*3*263+95693252643223 3**
        -     *   22 2 **22 2*+35673+9+587*786543522 53 *
        -             3*2*64453+56375+6++66582662323 3 *
       0+            ***22234*563955+84+868654953 2*322**
        -              *3*2**334356763778*8625353*34* * 2  *    *
        -      **        342 35273526376425546336*2*    *2  *
        -            *    * 2*2* 22333 453*2 2  2     **
        -         *   *    *2**2* 2 **2* *3 ** 2     * *
     -20+                    3         *2*  *  **
        -                        *     **
        -
          ----+---------+---------+---------+---------+---------+--X1
          -10.0      -5.0       0.0       5.0      10.0      15.0
```

M2.3.1 The Minitab commands and responses are

```
MTB > read 'b:\car.dat' c1-c4

   1242 ROWS READ

 ROW     C1     C2      C3      C4

   1      1    551   36400   12400
   2      2    661   15200   15400
   3      3    679   14100   16000
   4      4    561   22500   12100
  . . .

MTB > name c1 'I'
MTB > name c2 'Y'
MTB > name c3 'X1'
MTB > name c4 'X2'
```

M2.3.2 We use the `describe` command. The Minitab command and the response are

```
MTB > describe c2 c3

                N     MEAN   MEDIAN   TRMEAN    STDEV   SEMEAN
Y            1242   526.14   502.00   520.02   105.97     3.01
X1           1242    19648    19400    19497     5838      166

              MIN      MAX       Q1       Q3
Y          352.00   925.00   445.00   592.00
X1           7200    38300    15200    23625
```

From this we get $\mu_Y = \$526.14$, $\sigma_Y = \$105.97$, $\mu_{X1} = \$19,648$, and $\sigma_{X1} = \$5,838$.

M2.3.3 We use the `describe` command and find that the minimum value of $X1$ is \$7,200, the maximum value of $X1$ is \$38,300, the minimum value of $X2$ is 1,600 miles, and the maximum value of $X2$ is 18,500 miles.

M2.3.4 The command to print the Y data is `print c2`.

M2.3.5 The Minitab command and response are

```
MTB > histogram c2

Histogram of Y   N = 1242
Each * represents 10 obs.

Midpoint   Count
     350      32  ****
     400     147  ***************
     450     295  ******************************
     500     236  ************************
     550     185  *******************
     600     124  *************
     650      99  **********
     700      51  ******
     750      44  *****
     800      17  **
     850       9  *
     900       2  *
     950       1  *
```

M2.3.6 The Minitab commands and responses are

```
MTB > histogram c3

Histogram of X1   N = 1242
Each * represents 5 obs.

Midpoint   Count
    8000      14  ***
   10000      52  ***********
   12000      93  *******************
   14000     133  ***************************
   16000     158  ********************************
   18000     145  *****************************
   20000     146  ******************************
   22000     151  *******************************
   24000     108  **********************
   26000     102  *********************
   28000      61  *************
   30000      38  ********
   32000      22  *****
   34000      10  **
   36000       6  **
   38000       3  *

MTB > histogram c4

Histogram of X2   N = 1242
Each * represents 5 obs.

Midpoint   Count
    2000       1  *
    3000       7  **
    4000      18  ****
    5000      25  *****
    6000      39  ********
    7000      60  ************
    8000      92  *******************
    9000     122  *************************
   10000     146  ******************************
   11000     146  ******************************
   12000     157  ********************************
   13000     130  **************************
   14000     117  ************************
   15000      85  *****************
   16000      52  ***********
   17000      31  *******
   18000      12  ***
   19000       2  *
```

M2.3.7 The Minitab commands and response are

```
MTB > let k1= ssq(c2-mean(c2))
MTB > print k1
K1        13934922
```

So $SSY = 13934922$.

M2.3.8 The Minitab commands and response are

```
MTB > let k2=ssq(c2)
MTB > print k2
K2        357751808
```

So $\sum_{i=1}^{1242} Y_i^2 = 357751808$.

M2.3.9 Using the `describe` command we get $\hat{\sigma}_Y = \$105.97$, $\hat{\sigma}_{X1} = \$5,838$, and $\hat{\sigma}_{X2} = 3084.4$ miles.

M2.3.10 Using the `describe` command we find that the minimum value of Y is \$352.00 and the maximum value of Y is \$925.00.

M2.3.11 The Minitab command is

```
MTB > name c3 'X'
```

M2.3.12 The Minitab command and response are

```
MTB > plot c2 c3

         -
 Y       -                         *
         -              * *            *
         -      *          *     *  **   * *         *  *
     800+             3   *    *  *22    * *  **
         -       **  3  *  *2 *3  22* 222* 2*            *
         -        2 *  22 ***32*3***3**43 * 2  **
         -          **2*22**23*3 23433***22*4** 4 ***  3    **
         -      *3 *3*3443*2532252 3434*2 **4*252* 22 2  * *  *
     600+          222232443644343223584*3**4*2 2*    *  *      *
         -    * 2**4**474735+32+337363762326342 2*22* * 2     *  *
         -    ***3 425 464+7+327542347593625544**522   3*  *
         -     **536748538868+9+648++775368*53443*272  2*
         -    2*242423669399+5++4+5++956+97687*2544 **2* *  *  *
     400+      2 *234*47*73684228553235442244424*** * *2 2   *
         -   *     3***3**22223* *** * 2 2*  * *  * **
         -

          +---------+---------+---------+---------+---------+------X
        6000     12000     18000     24000     30000     36000
```

M2.3.13 The Minitab commands and response are

```
MTB > let k3=c2(792)
MTB > print k3
K3       528.000
```

So the first-year maintenance cost of car number 792 is \$528.00. You could, of course, use Table D-1 in Appendix D to find this number.

M2.3.14 We will print the data for the subpopulation of all cars that sold for \$12,500 using the following command.

```
MTB > copy c1 c2 c3 c4 c11 c12 c13 c14;
SUBC> use c3=12500.
MTB > print c11 c12 c13 c14
```

```
ROW    C11    C12    C13     C14

  1    292    484  12500   10800
  2    415    397  12500    7700
  3   1125    438  12500    8500
  4   1127    432  12500   10300
```

From these you can obtain the answers to (a), (b), and (c). There are 4 cars in the study population that sold for $12,500. They are the cars numbered 292, 415, 1125, and 1127. The first-year maintenance costs for these cars are $484, $397, $438, and $432, respectively. Using the `describe` command as shown below, we find that the mean first-year maintenance cost for these cars is $437.80 and the standard deviation is $35.70.

```
MTB > describe c12

              N     MEAN   MEDIAN   TRMEAN    STDEV   SEMEAN
C12           4    437.8    435.0    437.8     35.7     17.9

            MIN      MAX       Q1       Q3
C12       397.0    484.0    405.7    472.5
```

M2.3.15 We will obtain the mean first-year maintenance cost of all cars that sold for $9,600 with the following commands.

```
MTB > copy c2 c12;
SUBC> use c3=9600.
MTB > let k12=mean(c12)
MTB > print k12
K12       596.800
```

Since the best predicted value of the first-year maintenance cost of a car that sold for $9,600 is the average of the first-year maintenence costs of all cars that sold for that price, the anwer is $596.80.

M3.4.1 The Minitab command and response are

```
MTB > read 'b:\table323.dat' c1 c2
```

```
      26 ROWS READ

 ROW       C1     C2

   1       44      0
   2       86     10
   3       87     10
   4       58      3
  . . .
```

Use the following command to print columns c1 and c2.

```
MTB > print c1 c2
```

M3.4.2 The command is `regress c1 1 c2` . The Minitab response is

```
The regression equation is
C1 = 45.5 + 4.00 C2

Predictor        Coef        Stdev    t-ratio        p
Constant      45.5096       0.8080      56.33    0.000
C2             3.9979       0.1299      30.78    0.000

s = 2.817        R-sq = 97.5%      R-sq(adj) = 97.4%
```

From this we get $\hat{\beta}_0 = 45.5$, $\hat{\beta}_1 = 4.00$, $\hat{\mu}_Y(x) = 45.5 + 4.00x$, and $\hat{\sigma} = 2.817$.

M3.4.3 Use the following commands

```
MTB > brief 1
MTB > regress c1 1 c2

MTB > brief 2
MTB > regress c1 1 c2

MTB > brief 3
```

```
MTB > regress c1 1 c2

MTB > brief 0
MTB > regress c1 1 c2
```

M3.4.4 The Minitab command and response are

```
MTB > read 'b:\table324.dat' c1 c2

      10 ROWS READ

 ROW     C1     C2

    1    41      1
    2    59      4
    3    90     11
    4    88     11
  .   .   .
```

Print the data using the command `print c1 c2`. Then use the same commands as in **M3.4.2.** and **M3.4.3.**.

M3.4.5 The Minitab commands and responses are (the response to the `print` command and the `save` command are not shown)

```
MTB > read 'b:\arsenic.dat' c13 c21

      32 ROWS READ

 ROW       C1    C2

    1   0.17     0
    2   0.25     0
    3   0.01     0
    4   0.12     0
  .   .   .

MTB > print c13 c21
```

```
MTB > save 'c:t345';
MTB > portable.
```

M3.4.6 The Minitab command is

```
MTB > retrieve 'c:t345';
MTB > portable.
```

M3.4.7 The Minitab command and response are

```
MTB > regress c13 1 c21

The regression equation is
C13 = 0.105 + 0.988 C21

Predictor        Coef       Stdev    t-ratio        p
Constant      0.10458     0.06051       1.73    0.094
C21           0.98771     0.01446      68.29    0.000

s = 0.1875        R-sq = 99.4%      R-sq(adj) = 99.3%
```

From this we get $\hat{\beta}_0 = 0.105$, $\hat{\beta}_1 = 0.988$, $\hat{\mu}_Y(x) = 0.105 + 0.988x$, and $\hat{\sigma} = 0.1875$.

M3.5.1 The Minitab commands and responses are

```
MTB > retrieve 'b:\car20';
SUBC> portable.

Worksheet retrieved from file: b:\car20.MTP

MTB > info

COLUMN     NAME       COUNT
C1         mtcost        20
C2         miles         20

CONSTANTS USED: NONE
```

```
MTB > print c1 c2

 ROW  mtcost   miles

   1     456   11200
   2     828   17300
   3     500   11100
   4     489   11000
   5     387    6700
   6     553   13700
   7     531   12400
   8     650   15300
   9     475   11300
  10     474    8200
  11     533   12300
  12     396    7700
  13     618   14300
  14     474    8800
  15     639   13600
  16     457    7100
  17     460    8700
  18     433    6500
  19     621   13100
  20     460    9900
```

M3.5.2 Use the command `plot c1 c2`.

M3.5.3 The commands are

```
MTB > regress c1 1 c2  c3 c4;
MTB > residual c5.
MTB > name c3 'stdresid'
MTB > name c4 'fits'
MTB > name c5 'residual'
```

M3.5.4 From the Minitab output of Problem **M3.5.3** we get $\hat{\beta}_1 = 0.031307$, $\hat{\beta}_0 = 177.01$, $\hat{\mu}_Y(x) = 177.01 + 0.031307x$, and $\hat{\sigma} = 46.37$.

M3.5.5 From **M3.5.4** we have $\hat{\mu}_Y(x) = 177.01 + 0.031307x$. Substituting $x = 9400$ in this equation, we get $\hat{\mu}_Y(9400) = 471.3$.

M3.5.6 From the results of the following command you can obtain the desired point estimates.

```
MTB > describe c1 c2

                N      MEAN    MEDIAN    TRMEAN     STDEV    SEMEAN
mtcost         20     521.7     482.0     512.2     105.3      23.5
miles          20     11010     11150     10911      3038       679

              MIN       MAX        Q1        Q3
mtcost      387.0     828.0     457.7     601.7
miles        6500     17300      8325     13475
```

M3.6.1 First retrieve the file **arsenic.mtp** and examine its contents. Then, using the output from the following command you can obtain the answers needed.

```
MTB > regress c1 1 c2

The regression equation is
measured = 0.105 + 0.988 true

Predictor        Coef       Stdev    t-ratio        p
Constant      0.10458     0.06051       1.73    0.094
true          0.98771     0.01446      68.29    0.000

s = 0.1875          R-sq = 99.4%      R-sq(adj) = 99.3%
```

So $\hat{\beta}_0 = 0.10458$, $\hat{\beta}_1 = 0.98771$, $\hat{\mu}_Y(x) = 0.10458 + 0.98771x$, $\hat{\sigma} = 0.1875$, $SE(\hat{\beta}_0) = 0.06051$, and $SE(\hat{\beta}_1) = 0.01446$.

M3.6.2 Use the command `plot c1 c2`.

M3.6.3 The command and the subcommand are

```
MTB > regress c1 1 c2 c3 c4;
```

```
SUBC> residual c5.
```

M3.6.4 The required commands are

```
MTB > plot c3 c1
MTB > plot c3 c2
MTB > plot c3 c4
```

M3.6.5 This problem can be solved using the macro **confint**. The command is

```
execute 'b:\macro\confint'
```

The macro will ask you to input a number of values. The relevant part of the computer display is reproduced below.

```
*****************************************************************

  Input the column number for the column containing Y.
DATA> 1

  Input the column number for the column containing X.
DATA> 2

  Input the constants a(0) and a(1) in the linear function of
  interest in the indicated order.
DATA> 1 0

  Input the confidence coefficient for a two-sided confidence
  interval.
DATA> .90

=================================================================

  The estimated value of theta is    0.1045833

  The standard error of this estimate is   0.06050825
```

```
   For a two-sided confidence interval for theta with
                               confidence coefficient  0.900

   the lower confidence bound is  0.001885064   and

   the upper confidence bound is    0.2072816.
-----------------------------------------------------------------------
```

From this we get the confidence statement

$$C[0.00188506 \leq \beta_0 \leq 0.207282] = 0.90.$$

M3.6.6 You can again use the macro **confint** to solve this problem. The values to be input are the same as in problem **M3.6.5** except for the coefficients in the linear function of interest. We reproduce only this step below.

```
   Input the constants a(0) and a(1) in the linear function of
   interest, in the indicated order.
 DATA> 0 1
```

Note that we have used $a_0 = 0$ and $a_1 = 1$ corresponding to the parameter of interest which is β_1. The results displayed by the macro are given below.

```
=======================================================================

   The estimated value of theta is    0.9877083

   The standard error of this estimate is   0.01446424

   For a two-sided confidence interval for theta with
                               confidence coefficient  0.900

   the lower confidence bound is    0.9631587   and

   the upper confidence bound is     1.012258.
-----------------------------------------------------------------------
```

From this we get $C[0.963159 \leq \beta_1 \leq 1.01226] = 0.90$.

M3.6.7 From the results of the following command and subcommands you can obtain the answers to Problems **M3.6.7–M3.6.11**.

```
---------------------------------------------------------------------------
MTB > regress c1 1 c2;
SUBC> predict 3;
SUBC> predict 2.

The regression equation is
measured = 0.105 + 0.988 true

Predictor        Coef        Stdev     t-ratio        p
Constant      0.10458      0.06051        1.73    0.094
true          0.98771      0.01446       68.29    0.000

s = 0.1875       R-sq = 99.4%      R-sq(adj) = 99.3%

     Fit  Stdev.Fit         95% C.I.           95% P.I.
  3.0677     0.0339   ( 2.9984, 3.1370)  ( 2.6785, 3.4569)

  2.0800     0.0396   ( 1.9991, 2.1609)  ( 1.6886, 2.4714)
```

Thus $\hat{\mu}_Y(3) = 3.0677$.

M3.6.8 From the computer output in Problem **M3.6.7** above we get

$$C[2.9984 \leq \mu_Y(3) \leq 3.1370] = 0.95$$

M3.6.9 From the computer output in Problem **M3.6.7** we get

$$\hat{Y}(3) = \hat{\mu}_Y(3) = 3.0677$$

M3.6.10 From the computer output in Problem **M3.6.7** we get

$$C[2.6785 \leq Y(3) \leq 3.4569] = 0.95$$

M3.6.11 From the computer output in Problem **M3.6.7** we get

$$\hat{\mu}_Y(2) = 2.0800 \text{ and } SE(\hat{\mu}_Y(2)) = 0.0396$$

M3.6.12 An 85% confidence interval for $\mu_Y(2)$ can be obtained using the macro **confint** with $a_0 = 1$ and $a_1 = 2$ since the quantity of interest is $\mu_Y(2) = \beta_0 + 2\beta_1$. The relevant part of the macro input statements and the output from the macro are reproduced below.

```
*******************************************************************

  Input the column number for the column containing Y.
DATA> 1

  Input the column number for the column containing X.
DATA> 2

  Input the constants a(0) and a(1) in the linear function of
  interest in the indicated order.
DATA> 1 2

  Input the confidence coefficient for a two-sided confidence
  interval.
DATA> .85

===================================================================

  The estimated value of theta is     2.080000

  The standard error of this estimate is   0.03961195

  For a two-sided confidence interval for theta with
                           confidence coefficient  0.850

  the lower confidence bound is     2.021478   and

  the upper confidence bound is     2.138522.
-------------------------------------------------------------------
```

Thus we get $C[2.02148 \leq \mu_Y(2) \leq 2.13852] = 0.85$.

M3.7.1 We will use the macro **test**. You should first retrieve the file **crystal.mtp** and examine its contents. The command to invoke this macro is `execute 'b:\macro\test'`.

Minitab responds with:

```
--------------------------------------------------------------------------

  This program will compute the test statistic t_C for testing a
  linear combination

  theta=a(0) beta(0) + a(1) beta(1)

  equal to q, less than q, or greater than q for straight line
  regression. See Box 3.7.4 in the textbook.

 ************************************************************************

  Input the column number that contains the Y data.
DATA > 1

  Input the column number that contains the X data.
DATA > 2

  Input the values of a(0) and a(1) in this order.
DATA > 6 264

  Input the value of q.
DATA > 50

 ========================================================================

  The point estimate of theta is        132.914

  Its standard error is       6.35658

  The computed test statistic t_C is        13.0438

  If the test is NH: theta = q versus AH: theta not = q

     then the P-value is      0
```

```
  If the test is NH: theta < or = q versus AH: theta > q

     then the P-value is      0

  If the test is NH: theta > or = q versus AH: theta < q

     then the P-value is       1.000
--------------------------------------------------------------------
```

From this result the P-value is zero to the degree of accuracy of the arithmetic in the computer used. So the P-value is certainly less than 0.001.

M3.7.2 (a) Retrieve the file **shelflif.mtp** and examine its contents. Then use the macro **test** with $a_0 = 0$, $a_1 = 1$, and $q = 0$. The relevant portions of the computer display are reproduced below.

```
--------------------------------------------------------------------
  Input the values of a(0) and a(1) in this order.
DATA> 0 1

  Input the value of q.
DATA> 0

====================================================================

  The point estimate of theta is       -13.7534

  Its standard error is       0.774453

  The computed test statistic t_C is       -17.7588

  If the test is NH: theta = q versus AH: theta not = q

     then the P-value is               0

  If the test is NH: theta < or = q versus AH: theta > q

     then the P-value is        1.00000
```

```
  If the test is NH: theta > or = q versus AH: theta < q

     then the P-value is              0
--------------------------------------------------------------------------------
```

From this we observe that the *P*-value for the test is 0 (to machine accuracy).

(b) Since the quantity of interest is $\beta_0 + 13\beta_1$, we use the macro **test** with $a_0 = 1$, $a_1 = 13$. Also $q = 650$. The relevant portions of the results are reproduced below.

```
--------------------------------------------------------------------------------
  Input the values of a(0) and a(1) in this order.
DATA> 1 13

  Input the value of q.
DATA> 650

================================================================================

  The point estimate of theta is          746.959

  Its standard error is          7.71219

  The computed test statistic t_C is          12.5722

  If the test is NH: theta = q versus AH: theta not = q

     then the P-value is              0

  If the test is NH: theta < or = q versus AH: theta > q

     then the P-value is              0

  If the test is NH: theta > or = q versus AH: theta < q

     then the P-value is          1.00000
--------------------------------------------------------------------------------
```

Thus the *P*-value for the test is 1.0 (to machine accuracy).

M3.8.1 Retrieve the file **shelflif.mtp** and examine its contents. Then use the command `regress c1 1 c2` . As long as the value of **brief** is not set to zero, Minitab will print an analysis of variance table. Recall that the default value of **brief** is 2. The command and the Minitab response are given below.

```
MTB > regress c1 1 c2
```

Minitab responds with:

```
----------------------------------------------------------------------------

The regression equation is
days = 926 - 13.8 temp

Predictor        Coef        Stdev     t-ratio        p
Constant       925.75        17.13       54.05    0.000
temp         -13.7534       0.7745      -17.76    0.000

s = 17.05        R-sq = 95.2%      R-sq(adj) = 94.9%

Analysis of Variance

SOURCE        DF          SS          MS          F        p
Regression     1       91645       91645     315.38    0.000
Error         16        4649         291
Total         17       96295
----------------------------------------------------------------------------
```

M3.8.2 Retrieve the file **so2.mtp**. An ANOVA table is included in the output of the command `regress c1 1 c2` as long as the value of **brief** is not set to zero. The Minitab response is as follows.

```
----------------------------------------------------------------------------

The regression equation is
Y = 1.54 + 1.82 X

Predictor        Coef        Stdev     t-ratio        p
Constant        1.543        1.143        1.35    0.202
X              1.8248       0.2684        6.80    0.000
```

```
s = 1.514        R-sq = 79.4%      R-sq(adj) = 77.7%

Analysis of Variance

SOURCE        DF         SS          MS         F        p
Regression     1     105.96      105.96     46.23    0.000
Error         12      27.51        2.29
Total         13     133.47
-----------------------------------------------------------------
```

M3.8.3 First retrieve the file **grades26.mtp**. The command to obtain an ANOVA table is `regress c1 1 c2` with Y data in c1 and X data in c2. As long as the value of `brief` is not zero, Minitab will print an ANOVA table.

M3.11.1 Retrieve the file **gravity.mtp**. The command and the subcommand for a regression through the origin for these data are given below.

```
MTB > regress c1 1 c2;
SUBC >noconstant.
```

Minitab responds with:

```
-----------------------------------------------------------------

The regression equation is
ft/sec = 32.2 sec

Predictor        Coef        Stdev     t-ratio        p
Noconstant
sec           32.2071       0.0344      936.80    0.000

s = 0.8136

Analysis of Variance

SOURCE        DF         SS          MS         F        p
Regression     1     580888      580888 877600.56    0.000
Error          6          4           1
Total          7     580892
-----------------------------------------------------------------
```

M3.11.2 First retrieve the data from the file **gravity.mtp**. Then obtain a point estimate of β_1 and its standard error using the following commands.

```
regress c1 1 c2;
noconstant.
```

From the Minitab output we get $\hat{\beta}_1 = 32.2071$ and $SE(\hat{\beta}_1) = 0.0344$. To obtain a 90% confidence interval for β_1, we use the formula in (3.6.1). For this we also need the value of $t_{0.95:6}$. From Table T-2 we get $t_{0.95:6} = 1.943$. This t-value can also be obtained in Minitab using the following command and subcommand. You should consult the *Minitab Reference Manual* to learn more about these commands.

```
invcdf 0.95;
t 6.
```

Minitab responds with:

```
-----------------------------------------------------------------------

   0.9500     1.9432
-----------------------------------------------------------------------
```

Thus, using the formula in (3.6.1), the required interval is given by the statement

$$C[32.2071 - (1.9432)(0.0344) \leq \beta_1 \leq 32.2071 + (1.9432)(0.0344)] = 0.90$$

i.e.,

$$C[32.1403 \leq \beta_1 \leq 32.2739] = 0.90$$

M4.4.1 We retrieve the data file, examine its contents, and print the observations. The Minitab commands and responses are

```
MTB > retrieve 'b:\table444';
SUBC> portable.

Worksheet retrieved from file: table444.MTP
```

```
MTB > info

COLUMN     NAME       COUNT
C1         popitem       16
C2         strength      16
C3         temp          16
C4         pressure      16

CONSTANTS USED: NONE

MTB > print c1-c4

 ROW  popitem  strength   temp  pressure

   1     1150      36.6    260        10
   2     1186      20.7    230        18
   3      200      36.5    290        18
   4     1305      16.4    200        16
   5      783      23.2    200        10
   6     1066      26.6    230        14
   7     1023      22.5    210        16
   8      448      17.0    200        20
   9      945      32.7    290        18
  10      508      34.4    260        10
  11      704      32.4    260        12
  12     1135      24.8    240        18
  13      107      26.8    220        12
  14      742      37.7    280        12
  15      749      26.7    260        20
  16     1585      24.6    250        20
```

M4.4.2 First we form a column of 16 ones and put them in column c5 using the commands

```
MTB > set c5
DATA> 16(1)
DATA> end
```

Next we form the X matrix using the column of ones in c5 and the values of the predictor variables which are in columns c3 and c4. Then we print m1. The Minitab commands and response are

```
MTB > copy c5 c3 c4 m1
MTB > print m1

 MATRIX M1

    1   260    10
    1   230    18
    1   290    18
    1   200    16
    1   200    10
    1   230    14
    1   210    16
    1   200    20
    1   290    18
    1   260    10
    1   260    12
    1   240    18
    1   220    12
    1   280    12
    1   260    20
    1   250    20
```

Next we form the matrix y from the response variable Y, which is in column c2, and display it. The Minitab commands and response are

```
MTB > copy c2 m2
MTB > print m2

 MATRIX M2

  36.6
  20.7
  36.5
  16.4
  23.2
  26.6
  22.5
```

```
17.0
32.7
34.4
32.4
24.8
26.8
37.7
26.7
24.6
```

Next we compute X^T (which we label m3) and X^TX (which we label m4).

```
MTB > transpose m1 m3
MTB > multiply m3 m1 m4
MTB > print m4

 MATRIX M4

      16     3880     244
    3880   955400   59200
     244    59200    3936
```

Next we form $(X^TX)^{-1}$ and label it m5.

```
MTB > invert m4 m5
MTB > print m5

 MATRIX M5

  5.13008  -0.01658  -0.06862
 -0.01658   0.00007  -0.00001
 -0.06862  -0.00001   0.00465
```

Finally we compute X^Ty and label it m6.

```
MTB > multiply m3 m2 m6
MTB > print m6
```

```
MATRIX M6

     440
  109372
    6530
```

M4.4.3 $\hat{\beta} = (X^T X)^{-1} X^T y$; the Minitab commands and the response are

```
MTB > multiply m5 m6 m7
MTB > print m7

 MATRIX M7

 -6.52263
  0.19269
 -0.83479
```

M4.4.4 First we compute $X\hat{\beta}$, store it in m8, and then compute $y - X\hat{\beta}$, and store it in m9. The commands and response are

```
MTB > multiply m1 m7 m8
MTB > subtract m8 m2 m9
MTB > print m9

 MATRIX M9

  1.37036
 -2.07050
  2.16791
 -2.25930
 -0.46805
  0.49032
  1.91377
  1.67987
 -1.63209
 -0.82964
 -1.16005
```

```
 0.10256
 0.94767
 0.28609
-0.18171
-0.35478
```

M4.4.5 The required Minitab commands and the corresponding responses are

```
MTB > transpose m9 m10
MTB > multiply m10 m9 k1
ANSWER =         28.7602

MTB > let k2=k1/13
MTB > let k3=sqrt(k2)
MTB > print k3
K3        1.48739
```

Thus $\hat{\sigma}_{Y|X_1,X_2} = 1.48739$.

M4.4.6 Since Y is in c2 and X_1, X_2 are in c3 and c4 respectively, the command is `regress c2 2 c3 c4`. The Minitab response is (only the relevant portion of the output is shown)

```
The regression equation is
strength = - 6.52 + 0.193 temp - 0.835 pressure

Predictor       Coef       Stdev    t-ratio       p
Constant      -6.522       3.369      -1.94   0.075
temp         0.19269     0.01235      15.60   0.000
pressure     -0.8348      0.1015      -8.23   0.000

s = 1.487        R-sq = 95.9%     R-sq(adj) = 95.3%
```

M4.4.7 The Minitab commands are

```
MTB > set c5
DATA> 16(1)
```

```
DATA> end
MTB > copy c5 c3 m1
MTB > copy c2 m2
MTB > transpose m1 m3
MTB > multiply m3 m1 m4
MTB > invert m4 m5
MTB > multiply m3 m2 m6
MTB > multiply m5 m6 m7
MTB > print m7
MTB > multiply m1 m7 m8
MTB > subtract m8 m2 m9
MTB > transpose m9 m10
MTB > multiply m10 m9 k1
MTB > let k2=k1/14
MTB > let k3=sqrt(k2)
MTB > print k3
```

The values of $\hat{\beta}_0^A$ and $\hat{\beta}_1^A$ are the elements of the matrix m7, and they are equal to -18.8343 and 0.1910, respectively. The value of $\hat{\sigma}_{Y|X_1}$ is in **k3** and it is equal to 3.57118. Note that the degrees of freedom associated with $\hat{\sigma}_{Y|X_1}$ is 14 and so **k2** is obtained from k1 by dividing k1 by 14 (not 13 as in **M4.4.5**).

M4.4.8 Since Y is in column c2 and X_1 is in column c3 the command is `regr c2 1 c3`. The Minitab response is (only the relevant portion of the output is shown)

```
The regression equation is
strength = - 18.8 + 0.191 temp

Predictor         Coef        Stdev     t-ratio        p
Constant       -18.834        7.247       -2.60    0.021
temp           0.19097      0.02966        6.44    0.000

s = 3.571        R-sq = 74.8%      R-sq(adj) = 73.0%
```

So $\hat{\beta}_0^A = -18.834$, $\hat{\beta}_1^A = 0.19097$, and $\hat{\sigma}_{Y|X_1} = 3.571$.

M4.5.1 The required commands for **(a)**–**(f)** are given below.

```
MTB > retrieve 'b:\gpa';
```

```
SUBC> portable.
MTB > info
MTB > regr c1 4 c2 c3 c4 c5 c6 c7;
MTB > residual c8.
MTB > nscore c6 c9
MTB > name c6 'stdresid'
MTB > name c7 'fits'
MTB > name c8 'residual'
MTB > name c9 'nscores'
MTB > plot c6 c7
MTB > plot c6 c1
MTB > plot c6 c9
MTB > plot c6 c2
MTB > plot c6 c3
MTB > plot c6 c4
MTB > plot c6 c5
```

M4.5.2 Below we exhibit the commands for retrieving the data from the file **gpa.mtp**, examining their contents, and computing three different linear combinations of the variables for the **gpa** data.

```
MTB > retrieve 'b:\gpa';
SUBC> portable.

Worksheet retrieved from file: gpa.MTP

MTB > info

COLUMN     NAME       COUNT
C1         GPA           20
C2         SATmath       20
C3         SATverb       20
C4         HSmath        20
C5         HSengl        20

CONSTANTS USED: NONE


MTB > let c11=200*c1+c2-c3-200*c4
```

```
MTB > let c12=200*c1+c2+c3+200*c4+200*c5
MTB > let c13=200*c1-c2-c3-200*c4+200*c5
MTB > nscore c11 c21
MTB > nscore c12 c22
MTB > nscore c13 c23
```

The linear combinations above were chosen so that no single variable would dominate the values of the linear combination. Some care is needed here because the values taken by different variables are not commensurate with each other. For instance, Y, X_3, and X_4 all have values that range between 1 and 4 whereas X_1 and X_2 have values ranging between 200 and 800.

To obtain the rankit-plots, use the following commands.

```
MTB > plot c11 c21
MTB > plot c12 c22
MTB > plot c13 c23
```

You should try some other linear combinations.

M4.6.1 Valid point estimates for the specified quantities are available because the data are obtained by simple random sampling.

M4.6.2 The point estimates can be computed by first retrieving the **electric** data and then using the command `describe c1 c2 c3 c4`.

M4.6.3 These may be obtained by using the command **`regress c1 3 c2 c3 c4`** .

M4.6.4 The answer is σ. An estimate of σ is obtained from the computer output in Problem **M4.6.3**.

M4.6.5 The answer is σ_Y. An estimate of σ_Y is obtained from the Minitab output in Problem **M4.6.2**.

M4.6.6 The answer is β_2. The point estimate is 55.09; it can be obtained from the computer output for Problem **M4.6.3**. To obtain a 95% upper confidence bound, first obtain a 90% two-sided confidence interval for β_2 using the macro **cilinear** and then read off the upper bound. The relevant portions of the input and output from this macro are given below.

```
************************************************************************

  Input the column number for the column containing Y.
DATA> 1

  Input the column number for the column containing X(1).
DATA> 2

  Input the value of k, the number of predictors in your model.
DATA> 3

  Does your model have an intercept term? If yes, input 1, if no,
  input 0.
DATA> 1

  Input the constants a(0), a(1), a(2),...,a(k) in the linear
  function of interest, in the indicated order. If your model has
  no intercept, then input a(1),....,a(k) only.
DATA> 0 0 1 0

  Input the confidence coefficient for a two-sided confidence
  interval.
DATA> .90
========================================================================

  The estimated value of theta is      55.08763

  The standard error of this estimate is      29.04515

  For a two-sided confidence interval with
                        confidence coefficient      0.900

  the lower confidence bound is      5.790436   and

  the upper confidence bound is      104.3848
------------------------------------------------------------------------
```

So $C[\beta_2 \leq 104.3848] = 0.95$.

M4.6.7 The quantity of interest is $1000\beta_1$. As in Problem **M4.6.6**, use the macro **cilinear**, but now specify $a_0 = 0, a_1 = 1000, a_2 = 0$, and $a_3 = 0$. Specify the confidence coefficient as 0.90 for a two-sided interval. From the result read off the upper bound.

M4.6.8 The quantity of interest is $500\beta_3$. Use the macro **cilinear** with $a_0 = 0, a_1 = 0, a_2 = 0$, and $a_3 = 500$. Specify the confidence coefficient as 0.90 for a two-sided interval. From the result read off the upper bound.

M4.6.9–M4.6.11 Use the command and subcommand

```
regress c1 3 c2 c3 c4;
predict 3200 6 2800.
```

Minitab will give a point estimate and a two-sided 95% confidence interval for the quantities $\mu_Y(3200, 6, 2800)$ and $Y(3200, 6, 2800)$. To obtain a 90% upper confidence bound for $\mu_Y(3200, 6, 2800)$, use the macro **cilinear** with $a_0 = 1$, $a_1 = 3200$, $a_2 = 6$, and $a_3 = 2800$, get an 80% two-sided interval, and use only the upper bound. To obtain a 90% upper confidence bound for $Y(3200, 6, 2800)$, you can use the macro **pred** discussed in Chapter 6, get an 80% two-sided interval, and read off the upper bound.

M4.6.12 The quantity of interest is $\theta = 400\beta_1 + 3\beta_2 + 1100\beta_3$. An estimate and a confidence interval for this quantity can be obtained using the macro **cilinear** with $a_0 = 0, a_1 = 400, a_2 = 3$, $a_3 = 1100$, and $1 - \alpha = 0.95$. We get $\hat{\theta} = \$504.53$ and $C[\$237.40 \leq \theta \leq \$771.67] = 0.95$.

M4.6.13 The answer can be obtained using the macro **sgmaconf**. First you need the value of $\hat{\sigma}$ and its associated degrees of freedom. These are 135.4 and 30 respectively, obtained using the command `regress c1 3 c2 c3 c4` . The relevant portions of the Minitab response from the macro **sgmaconf** are given below.

```
*******************************************************************

   Input the estimate of sigma_Y or sigma, whichever applies
 DATA> 135.4

   Input the associated degrees of freedom
 DATA> 30

   Input the confidence coefficient 1-alpha
 DATA> .95

===================================================================
```

```
For a two-sided  95.0% confidence interval for sigma

the lower confidence bound is        108.200     and

the upper confidence bound is        180.986
----------------------------------------------------------------------
```

M4.7.1 **(a)** First retrieve the file **gpa.mtp**. Then use the macro **testmult**. The relevant portions of the input and output for this macro are given below.

```
*********************************************************************

  Input the column number for the column containing Y.
DATA> 1

  Input the column number for the column containing X(1).
DATA> 2

  Input k, the number of predictors in your model.
DATA> 4

  Does your model have an intercept term. If yes, input 1, if no,
  input 0.
DATA> 1

  Input the constants a(0), a(1), a(2), ..., a(k) in the linear
  function of interest, in this order. If your model has no
  intercept, then only input a(1), ..., a(k).
DATA> 0 1 0 0 0

  Input the value of q.
DATA> .003
=====================================================================

  The estimated value of theta is  0.002010167

  The standard error of this estimate is  0.000584442

  The test statistic t_C is    -1.693639

  If the test is NH: theta = q against AH: theta not equal to q
  then the P-value is    0.1109942
```

```
If the test is NH: theta < or = q against AH: theta > q
then the P-value is    0.9445029

If the test is NH: theta > or = q against AH: theta < q
then the P-value is   0.05549711
-----------------------------------------------------------------------
```

Thus the required P-value is 0.111 (rounded to 3 decimals). So do not reject NH.

(b) Use the macro **testmult** with $a_0 = a_1 = a_3 = a_4 = 0$ and $a_2 = 1$. Also the value of q is 0.001. The P-value is 0.327, so do not reject NH.

(c) Use the macro **testmult** with $a_0 = 1$, $a_1 = 500$, $a_2 = 615$, $a_3 = 3.10$, $a_4 = 2.90$ and $q = 2.5$. The P-value is 0.007, so reject NH.

M4.7.2 The hypothesis of interest is NH: $\beta_1 = 0$ against AH: $\beta_1 \neq 0$. Retrieve the file **electric.mtp** and use the macro **testmult** with $a_0 = 0, a_1 = 1, a_2 = 0, a_3 = 0$, and $q = 0$. The P-value is 0.585. Do not reject NH.

M4.7.3 Retrieve the file **electric.mtp** and use the macro **sgmatest**. The relevant portions of the input and output from this macro are given below.

```
***********************************************************************

  Input the estimate of sigma
DATA> 135.4

  Input the degrees of freedom associated with the estimate of sigma
DATA> 30

  Input the value of q
DATA> 50

=======================================================================

  The test statistic Chisquare_C is     219.9979

  For a test of NH: sigma = q against AH: sigma not = q

                              the P-value is            0
```

```
For a test of NH: sigma < or = q against AH: sigma > q

                                    the P-value is          0

For a test of NH: sigma > or = q against AH: sigma < q

                                    the P-value is     1.000000
-----------------------------------------------------------------------
```

The P-value is 1.00 (to machine precision), so do not reject NH.

M4.9.1 First retrieve the file **age18.mtp**. From Exhibit 4.9.2 we get $SSE(A) = 12.08810$ with 12 degrees of freedom, and $SSE(B) = 13.85183$ with 16 degrees of freedom. These may also be obtained by using the `regress` command. We execute the macro **ratiosgm** and show the relevant parts of the input and the output below.

```
*******************************************************************

 Input the quantity SSE(A), the sum of squared errors for Model-A.
DATA > 12.08810

 Input the degrees of freedom associated with the sum of squared
 errors for Model-A.
DATA > 12

 Input the confidence coefficient for sigma(A)
DATA > .80

 Input the quantity SSE(B), the sum of squared errors for Model-B.
DATA > 13.85183

 Input the degrees of freedom associated with the sum of squared
 errors for Model-B.
DATA > 16

 Input the confidence coefficient for sigma(B)
DATA > .80
=================================================================

 The estimate of sigma(A) is     1.003664
```

```
For a two-sided confidence interval for sigma(A)
                    with confidence coefficient  0.800
the lower bound is    0.8072624   and
the upper bound is     1.384771.

The estimate of sigma(B) is    0.9304512

For a two-sided confidence interval for sigma(B)
                    with confidence coefficient  0.800
the lower bound is    0.7670671   and
the upper bound is     1.219626.

The estimate of sigma(B)/sigma(A) is    0.9270543

For a two-sided confidence interval for sigma(B)/sigma(A)
with confidence coefficient greater than or equal to  0.600
the lower bound is    0.5539306   and
the upper bound is     1.510817.
```

From this we observe

$$C[0.8072624 \leq \sigma_A \leq 1.384771] = 0.80,$$

$$C[0.7670671 \leq \sigma_B \leq 1.219626] = 0.80,$$

and $\frac{\hat{\sigma}_B}{\hat{\sigma}_A} = 0.9271$. You can also obtain the confidence intervals for σ_A and σ_B using the macro **sgmaconf**.

M4.9.2 For this problem, use the macro **ratiosgm** with confidence coefficients for Model-A and Model-B each equal to $1 - (1 - 0.95)/2 = .975$. The relevant portions of the input and output from the macro are given below.

```
*****************************************************************

  Input the quantity SSE(A), the sum of squared errors for Model-A.
DATA> 12.08810

  Input the degrees of freedom associated with the sum of squared
  errors for Model-A.
DATA> 12

  Input the confidence coefficient for sigma(A)
DATA> .975
```

```
  Input the quantity SSE(B), the sum of squared errors for Model-B.
DATA> 13.85183

  Input the degrees of freedom associated with the sum of squared
  errors for Model-B.
DATA> 16

  Input the confidence coefficient for sigma(B)
DATA> .975

=======================================================================

  The estimate of sigma(A) is     1.003664

  For a two-sided confidence interval for sigma(A)
                      with confidence coefficient  0.975
  the lower bound is    0.6880974   and
  the upper bound is     1.794816.

  The estimate of sigma(B) is    0.9304512

  For a two-sided confidence interval for sigma(B)
                      with confidence coefficient  0.975
  the lower bound is    0.6657745   and
  the upper bound is     1.512611.

  The estimate of sigma(B)/sigma(A) is    0.9270543

  For a two-sided confidence interval for sigma(B)/sigma(A)
  with confidence coefficient greater than or equal to  0.950
  the lower bound is    0.3709430   and
  the upper bound is     2.198251.
-----------------------------------------------------------------------
```

From this we get $C[0.3709 \leq \sigma_B/\sigma_A \leq 2.1983] \geq 0.95$.

M4.11.1 First retrieve the data from the file **car17.mtp**. Then use the macro **lackfit** with $1 - \alpha = 0.95$.

M5.2.1 First retrieve the data from the file **premiums.mtp**. Then change the Y value for sample item number 36 from 491 to 1491 with the command

```
MTB > let c1(36)=1491
```

and give the following **regress** command and subcommands

```
MTB > regress c1 2 c2 c3 c4 c5;
MTB > residuals c6;
MTB > tresiduals c7.
```

Finally name the variables and print the data. Note that the standardized residuals are in c4, fits are in c5, residuals are in c6, and the studentized deleted residuals are in c7.

M5.4.1 The commands to enter the data via the keyboard, name them, and print them, and the corresponding Minitab responses are as follows.

```
MTB > set c1
DATA> 8.8 3.1 8.5 15.3 15.2 16.4 21.5 22.5 22.3 52.2
MTB > set c2
DATA> 1.0 1.5 2.0 2.5 3.0 3.5 4.0 4.5 5.0 11.0
DATA> end
MTB > name c1 'Y'
MTB > name c2 'X'
MTB > print c1 c2

 ROW      Y      X

   1    8.8    1.0
   2    3.1    1.5
   3    8.5    2.0
   4   15.3    2.5
   5   15.2    3.0
   6   16.4    3.5
   7   21.5    4.0
   8   22.5    4.5
   9   22.3    5.0
  10   52.2   11.0
```

The commands below are used to compute the regression of Y on X and obtain the standardized residuals, the fits, the residuals, the studentized deleted residuals, the

Cook's distances, the DFFITS, and the hat values. Note the use of the subcommand `predict 4.5` for obtaining $\hat{Y}(4.5)$ in part (d).

```
MTB > regress c1 1 c2 c3 c4;
SUBC> residuals c5;
SUBC> tresiduals c6;
SUBC> cook c7;
SUBC> dfit c8;
SUBC> hi c9;
SUBC> predict 4.5.
```

Minitab responds with:

```
-------------------------------------------------------------------------

The regression equation is
Y = 0.86 + 4.66 X

Predictor        Coef        Stdev     t-ratio        p
Constant        0.857        1.401        0.61    0.558
X              4.6640       0.3008       15.50    0.000

s = 2.563       R-sq = 96.8%     R-sq(adj) = 96.4%

Analysis of Variance

SOURCE        DF          SS          MS          F        p
Regression     1      1579.3      1579.3     240.37    0.000
Error          8        52.6         6.6
Total          9      1631.9

     Fit  Stdev.Fit         95% C.I.            95% P.I.
  21.845      0.847   ( 19.913, 23.777)  ( 15.625, 28.065)

-------------------------------------------------------------------------
```

Next we name the newly created columns and print them.

```
MTB > name c3 'stdresid'
MTB > name c4 'fits'
```

```
MTB > name c5 'residual'
MTB > name c6 'tresid'
MTB > name c7 'cooksd'
MTB > name c8 'dffits'
MTB > name c9 'hatvals'
MTB > print c1-c5

 ROW      Y       X  stdresid      fits  residual

   1    8.8     1.0   1.43757    5.5207   3.27934
   2    3.1     1.5  -2.03873    7.8527  -4.75269
   3    8.5     2.0  -0.71065   10.1847  -1.68471
   4   15.3     2.5   1.15966   12.5167   2.78327
   5   15.2     3.0   0.14515   14.8488   0.35124
   6   16.4     3.5  -0.32131   17.1808  -0.78078
   7   21.5     4.0   0.81745   19.5128   1.98719
   8   22.5     4.5   0.27044   21.8448   0.65516
   9   22.3     5.0  -0.78047   24.1769  -1.87686
  10   52.2    11.0   0.03514   52.1612   0.03885

MTB > print c6-c9

 ROW     tresid      cooksd      dffits     hatvals

   1    1.56144    0.271355    0.80017    0.207989
   2   -2.75131    0.434329   -1.25778    0.172865
   3   -0.68678    0.042695   -0.28240    0.144628
   4    1.18933    0.094549    0.44598    0.123278
   5    0.13596    0.001286    0.04751    0.108815
   6   -0.30251    0.005814   -0.10153    0.101240
   7    0.79874    0.037351    0.26706    0.100551
   8    0.25414    0.004370    0.08785    0.106749
   9   -0.75956    0.041467   -0.28026    0.119835
  10    0.03288    0.002703    0.06879    0.814050
```

From this output you can obtain the answers for **(a)** to **(g)**. For example, for **(c)** we read Cook's distance for observation 9 and get $c_9 = 0.041467$. Likewise, for **(a)** we get $h_{4,4} = 0.123278$. For part **(d)**, the value of $\hat{Y}(4.5)$ is 21.845.

M6.2.1 **(a)** The required point estimates are obtained by using the following commands.

```
regress c1 3 c2 c3 c4;
predict 10 20 30.2;
predict 8.5 15 15;
predict 6.5 24 28.
```

From the output we get the following answers for (i)–(iii).

(i) \$149.49.

(ii) \$47.81.

(iii) \$56.10.

(b) Use the macro **pred** to solve this problem. You must input the following numbers in c12 using the `set` command.

```
set c12
1 10 20 30.2
end
```

Then type `execute 'b:\macro\pred'`. The relevant input and output statements are displayed below.

```
*************************************************************************

  Input the value of k, the number of predictor variables (note
  that k must be less than or equal to 9)
DATA> 3

  Does your model have an intercept term? If yes, enter 1. If no,
  enter 0.
DATA> 1

  Input h, the number of future Y values under consideration.
DATA> 1

  Input the value of the confidence coefficient 1-alpha
DATA> .90

=========================================================================
```

```
The estimate of the average, Y_A, of the h future Y values is

YAhat =      149.4865


The standard error of this estimate is     55.42014


For a two-sided prediction interval for the average Y_A of the
h future Y values, with confidence coefficient  0.900
the lower bound is      56.05061 and
the upper bound is      242.9223
-----------------------------------------------------------------
```

Note that, when $h = 1$, the average Y_A and the sum Y_S are the same and each is equal to the Y value to be predicted. So, we have shown only the results for Y_A in the output above. Thus we have

$$C[56.05 \leq Y(10, 20, 30.2) \leq 242.92] = 0.90.$$

(c) You can use the macro **pred** to solve this problem. First you must type in the appropriate X values for the subpopulations from which the future Y values will be chosen. This is done as follows.

```
set c12
1 10 20 30.2
end
set c13
1 8.5 15 15
end
set c14
1 6.5 24 28
end
```

Then you invoke the macro **pred** by typing `execute 'b:\macro\pred'` . The relevant portions from the input and the output statements are displayed below.

```
*****************************************************************
```

```
  Input the value of k, the number of predictor variables (note
  that k must be less than or equal to 9)
DATA> 3

  Does your model have an intercept term? If yes, enter 1. If no,
  enter 0.
DATA> 1

  Input h, the number of future Y values under consideration.
DATA> 3

  Input the value of the confidence coefficient 1-alpha
DATA> .90
========================================================================

  The estimate of the average, Y_A, of the h future Y values is

  YAhat =      84.46390

  The standard error of this estimate is      37.55078

  For a two-sided prediction interval for the average Y_A of the
  h future Y values, with confidence coefficient  0.900
  the lower bound is     21.15498 and
  the upper bound is     147.7728


  The estimate of the sum, Y_S, of the h future Y values is

  YShat =      253.3917

  The standard error of this estimate is      112.6523

  For a two-sided prediction interval for the sum Y_S of the
  h future Y values, with confidence coefficient  0.900
  the lower bound is     63.46495 and
  the upper bound is     443.3184

------------------------------------------------------------------------
```

Hence we get $\hat{Y}_S$ = \$253.39, the standard error of this estimate is \$112.65, and the confidence statement is

$$C[\$63.46 \leq Y_S \leq \$443.32] = 0.90.$$

M6.2.2 For parts (c), (d), and (e) of Exercise **6.9.1**, the answers can be obtained using the macro **pred**.

(c) First retrieve the file **track.mtp** and then enter the appropriate X values in c12 using the following commands.

```
retrieve 'b:\track';
portable.
set c12
1 52.08
end
```

Then execute the macro **pred**. The relevant portions of the input and output statements are displayed below (we use $1 - \alpha = 0.90$ to get a two-sided prediction interval from which we can get a lower bound with confidence coefficient 0.95).

```
*********************************************************************

  Input the value of k, the number of predictor variables (note
  that k must be less than or equal to 9)
DATA> 1

  Does your model have an intercept term? If yes, enter 1. If no,
  enter 0.
DATA> 1

  Input h, the number of future Y values under consideration.
DATA> 1

  Input the value of the confidence coefficient 1-alpha
DATA> .90
=====================================================================

  The estimate of the average, Y_A, of the h future Y values is

  YAhat =      51.07378
```

```
The standard error of this estimate is     0.4188021


For a two-sided prediction interval for the average Y_A of the
h future Y values, with confidence coefficient  0.900
the lower bound is       50.32166 and
the upper bound is       51.82591.

The estimate of the sum, Y_S, of the h future Y values is

YShat =       51.07378

The standard error of this estimate is     0.4188021

For a two-sided prediction interval for the sum Y_S of the
h future Y values, with confidence coefficient  0.900
the lower bound is       50.32166 and
the upper bound is       51.82591.
---------------------------------------------------------------
```

Thus the predicted race time for this athlete is 51.07 seconds. To obtain a 95% lower prediction bound we take the lower bound of the 90% two-sided prediction interval from the macro output. Hence $C[50.32 \leq Y(52.08)] = 0.95$.

(d) Create columns `c12, c13, c14, c15` as follows.

```
set c12
1 51.32
end
set c13
1 52.07
end
set c14
1 52.58
end
set c15
1 51.96
end
```

Then execute the macro **pred** with $k = 1$, $h = 4$, and $1 - \alpha = 0.98$. The predicted total time for the team is 203.896 seconds. A 99% lower confidence bound for the total team time Y_S is given by the lower bound from a 98% two-sided prediction interval for Y_S. Hence we get

$$C[201.32 \leq Y_S] = 0.99.$$

(e) Repeat the procedure from **(d)** above but now use $1 - \alpha = 0.95$. The answer is

$$C[201.81 \leq Y_S \leq 205.98] = 0.95.$$

M6.3.1 Relevant portions of the input and output statements from the macro **toleranc** are given below.

```
**********************************************************************

  Input the value of k, the number of predictor variables (note k
  must be less than or equal to 9)
DATA> 1

  Does your model contain an intercept term? If yes, input 1. If no,
  input 0.
DATA> 1

  Input the  x-vector corresponding to the subpopulation for which
  a point estimate and a confidence interval for a tolerance point
  are required. Remember, if the model has an intercept term, then
  the x-vector contains k+1 elements, one for the intercept term and
  one for each predictor variable. In particular, the first element is
  the number 1 in this case. If the model does not contain an intercept
  term, then the x-vector contains k elements, one for each predictor
  variable.
DATA> 1 3

  Input the value of p.
DATA> .20

  Input confidence coefficient 1-alpha for a two-sided tolerance
  interval.
DATA> .95
```

```
  Input the noncentral t-value in (8) of Box 6.3.1.
  This value is obtained from Table T-8.
  The value of alpha/2 to use is 0.02500.
  The degrees of freedom are    16 and
  the value of delta is     3.241651

DATA> 1.254

  Input the noncentral t-value in (8) of Box 6.3.1.
  This value is obtained from Table T-8.
  The value of 1-alpha/2 to use is  0.9750.
  The degrees of freedom are    16 and
  the value of delta is     3.241651

DATA> 6.174
=======================================================================

  The estimate of the tolerance point is   -1.113384


  A two-sided    95% confidence interval for the p-th tolerance
  point, where p =    0.200, has
  lower bound =     -1.916568    and
  upper bound =    -0.5689563
-----------------------------------------------------------------------
```

Thus we have

$$C[-1.92 \leq \lambda_{0.20}(3.0) \leq -0.57] = 0.95.$$

M6.3.2 (a) Use the macro **toleranc** with $t_{0.025:18;(-9.112)} = -14.119$ and $t_{0.975:18;(-9.112)} = -6.319$.

Minitab responds with:

```
-----------------------------------------------------------------------

  The estimate of the tolerance point is    149.4056
```

```
A two-sided    95% confidence interval for the p-th tolerance
point, where p =  0.990, has
lower bound =      145.2056   and
upper bound =      156.9351
------------------------------------------------------------------------
```

(b) What is needed here is an 80% confidence interval for $\lambda_{0.95}(240)$. Use the macro **toleranc** with $t_{0.10:18;(-4.469)} = -6.416$ and $t_{0.90:18;(-4.469)} = -3.072$.

Minitab responds with:

```
------------------------------------------------------------------------

The estimate of the tolerance point is    157.9589


A two-sided    80% confidence interval for the p-th tolerance
point, where p =  0.950, has
lower bound =      154.9310   and
upper bound =      162.1810
------------------------------------------------------------------------
```

Based on this result, one might conclude that a blood pressure value of 210 units is in the upper 5% of the population of blood pressures of individuals who weigh 240 pounds.

(c) Use the macro **toleranc** with $t_{0.025:18;(-8.483)} = -13.227$ and $t_{0.975:18;(-8.483)} = -5.808$.

Minitab responds with:

```
------------------------------------------------------------------------

The estimate of the tolerance point is    128.4599


A two-sided    95% confidence interval for the p-th tolerance
point, where p =  0.990, has
lower bound =      124.1387   and
upper bound =      136.1218
------------------------------------------------------------------------
```

M6.4.1 For Problems 6.4.1 and 6.4.2, we use the macro **regul** (since we want to regulate the average temperature) with $1-\alpha = 0.99$ and $\mu_Y(x_0) =$ `mu(x0)` $= 400$, and get $\hat{x}_0 = 66.52$ and $C[65.071 \le x_0 \le 68.026] = 0.99$.

For Problems 6.4.3 and 6.4.4 we use the macro **calib** with $1-\alpha = 0.90$ and $y_0 = 100$, and get $\hat{x}_0 = 100.1$ and $C[99.6 \le x_0 \le 100.6] = 0.90$.

For Problems 6.4.5 and 6.4.6 we use the macro **regul** with $1-\alpha = 0.90$ and $\mu_Y(x_0) =$ `mu(x0)` $= 5$, and we get $\hat{x}_0 = 9.929$ hours and $C[8.650 \le x_0 \le 11.048] = 0.90$.

M6.4.2 Using the macro **regul** with $1-\alpha = 0.90$ and $\mu_Y(x_0) =$ `mu(x0)` $= 15$, we get $\hat{x}_0 = 16.862$ and $C[15.989 < x_0 \le 17.954] = 0.90$.

M6.5.1 First retrieve the file **eggshell.mtp** and then use the macro **compare.** Below we list relevant portions of the input and the output from the macro.

```
*********************************************************************

  Input the number of regression lines to be compared.
DATA> 3

  Input the number of intervals desired.
DATA> 3

  Input the simultaneous confidence coefficient desired.
DATA> .90

...

...

  Input the coefficients for linear function    1

DATA> 1 0 -1 0 0 0

  Input the coefficients for linear function    2

DATA> 1 0 0 0 -1 0

  Input the coefficients for linear function    3

DATA> 0 0 1 0 -1 0
=====================================================================
```

```
The pooled estimate of the subpopulation standard deviation
is      1.217584   and the associated degrees of freedom =   23


The required simultaneous confidence intervals (Bonferroni)
with simultaneous confidence coefficient greater than or
equal to   0.900 are given below

Linear Function   Point Estimate    Lower Bound    Upper Bound


        1            -0.5011616       -3.468777       2.466454
        2             0.9436436       -1.813539       3.700826
        3              1.444805       -1.554462       4.444073
```

From the preceeding output you can obtain the desired confidence intervals.

M6.6.1 First retrieve the file **eggshell.mtp** and then use the macro **inter**. The relevant portions of the input and output statements from the macro are given below.

```
************************************************************************

  Input the number of the column containing the Y values
  for the first regression line.
DATA> 1

  Input the number of the column containing the X values
  for the first regression line.
DATA> 2

  Input the number of the column containing the Y values
  for the second regression line.
DATA> 5

  Input the number of the column containing the X values
  for the second regression line.
DATA> 6
```

```
 Input the confidence coefficient desired.
DATA> .90

==========================================================================

 The point estimate for the point of intersection is    -0.3387100


 The confidence region with confidence coefficient equal to  0.900
 is in fact an interval.

 The lower bound is     -1.247561   and
 the upper bound is     0.4485041

--------------------------------------------------------------------------
```

From the output we get $\hat{x}_0 = -0.339$ and $C[-1.248 \leq x_0 \leq 0.449] = 0.90$.

M6.6.2 First retrieve the data from the file **eggshell.mtp**. Then use the macro **inter** by issuing the command `execute b:\macro\inter'` . The relevant portions of the input and output statements from the macro are given below.

```
 Input the confidence coefficient desired.

DATA> .90

==========================================================================

 The point estimate for the point of intersection is     0.2564005


 The confidence region with confidence coefficient equal to  0.900
 is in fact an interval.

 The lower bound is     -1.027576   and
 the upper bound is      1.233465

--------------------------------------------------------------------------
```

So we have $\hat{x}_0 = 0.256$ and $C[-1.028 \leq x_0 \leq 1.233] = 0.90$.

M6.7.1 Retrieve the file **concrete.mtp** and use the macro **quadr** with $1-\alpha = 0.90$. The relevant portions of the input and output from the macro are displayed below.

```
*****************************************************************

 Input the confidence coefficient desired.
DATA> .90

=================================================================

  The estimate of the point of extremum is      31.82399


  The confidence region with confidence coefficient equal to  0.900
  is in fact an interval.

  The lower bound is      29.90076  and
  the upper bound is      33.97816

-----------------------------------------------------------------
```

So $\hat{x}_0 = 31.824$ and $C[29.901 \leq x_0 \leq 33.978] = 0.90$.

M6.8.1 First retrieve the data from the file **sales.mtp**. Answers to parts **(a)**, **(b)**, and **(d)** are obtained by using the macro **spline**. The relevant portions of the input and the output from this macro are shown below.

```
*****************************************************************

 Input the knot point q.
DATA> 50

 Input the number of linear functions for which confidence
 intervals are required.
DATA> 5

 Input the coefficients for the linear functions of interest.
 Note that each linear function of interest can be written in
 the form

   a(1) alpha(1) + b(1) beta(1) + a(2) alpha(2) + b(2) beta(2).

 The coefficients to be input are

   a(1)  b(1)  a(2)  b(2)

 in this order.
```

```
DATA> 1 0 0 0
DATA> 0 1 0 0
DATA> 0 0 1 0
DATA> 0 0 0 1
DATA> 0 0 1 75
     5 ROWS READ

  Input the confidence coefficient for two-sided one-at-a-time
  confidence intervals for each linear function.
DATA> .90
==================================================================

  Point estimates for the intercepts and slopes of the two lines
  are given below. Line 1 is the line to the left of the knot point
  and line 2 is to the right of the knot point.

  Estimated intercept for line 1 is     201.4454
  Estimated   slope   for line 1 is     5.021784

  Estimated intercept for line 2 is     404.2462
  Estimated   slope   for line 2 is    0.9657681


  Estimated subpopulation standard deviation is      11.04885

 The point estimates and corresponding standard errors for the
 linear functions of interest are given below. Also given are
 lower bounds and upper bounds for two-sided confidence intervals
 with one-at-a-time confidence coefficient equal to  0.900

  Linear        Point      Standard          Lower          Upper
 Function     Estimate        Error          Bound          Bound


        1     201.4454     11.69917       181.0933       221.7974
        2     5.021784    0.2874708       4.521696       5.521873
        3     404.2462     15.27978       377.6653       430.8271
        4    0.9657681    0.2398292      0.5485577       1.382978
        5     476.6788     4.303996       469.1915       484.1661
------------------------------------------------------------------
```

To answer **(c)**, you must first create a column consisting of the values of $\hat{\mu}_Y(x_i)$ using the spline model. The command to do this is

```
MTB > let c3=(c2<=50)*(201.4454+5.021784*c2)+(c2>50)*(404.2462+0.9657681*c2)
```

This command computes $\hat{\mu}_Y(x_i)$ by using the equation $\hat{\mu}_Y(x_i) = 201.4454 + 5.021784x_i$ for $x_i \leq 50$, and $\hat{\mu}_Y(x_i) = 404.2462 + 0.9657681x_i$ for $x_i > 50$. The result is stored in c3. We name c3 and plot $\hat{\mu}_Y(x_i)$ in c3 against x_i in c2 using the following commands.

```
MTB > name c3 'fits'
MTB > plot c3 c2
```

Minitab responds with:

```
-----------------------------------------------------------------------------


          -                                                       *
      480+                                              2     *
          -                                  * *   **
fits      -
          -                         *3
          -                        *
      400+                        3
          -                   *
          -
          -                *
          -
      320+             *
          -
          -
          -
          -    *
      240+
           ------+---------+---------+---------+---------+---------+advbudgt
                15        30        45        60        75        90
-----------------------------------------------------------------------------
```

M7.4.1 (a) You must first retrieve the data from the file **table742.mtp**. For stepwise regression with *F-in* = `fenter` = 3.0 and *F-out* = `fremove` = 3.0, the Minitab command and subcommands along with the output are given below.

Minitab responds with:

```
MTB > stepwise c1 c2-c4;
SUBC> fenter=3.0;
SUBC> fremove=3.0.

 STEPWISE REGRESSION OF     Y      ON  3 PREDICTORS, WITH N =   10

    STEP         1         2         3         4
CONSTANT    9.8728    7.6917    1.7361    0.2626

X1            0.33      0.47      0.19
T-RATIO       2.16      3.04      1.01

X2                      0.60      1.55      1.80
T-RATIO                 1.83      2.97      3.85

X3                                1.14      1.54
T-RATIO                           2.12      4.15

S            0.819     0.720     0.588     0.589
R-SQ         36.88     57.32     75.60     71.41
 MORE? (YES, NO, SUBCOMMAND, OR HELP)
SUBC> yes

 NO VARIABLES ENTERED OR REMOVED

 MORE? (YES, NO, SUBCOMMAND, OR HELP)
SUBC> no
```

Hence the final model chosen by this procedure uses X_2 and X_3 as the predictor variables.

(b) For forward selection with *F-in* = `fenter` = 3.0, the Minitab command, subcommands, and output are as follows.

```
MTB > stepwise c1 c2-c4;
SUBC> fenter=3.0;
SUBC> fremove=0.0.
```

Minitab responds with:

```
--------------------------------------------------------------------------

 STEPWISE REGRESSION OF     Y     ON  3 PREDICTORS, WITH N =   10

    STEP         1        2        3
CONSTANT     9.873    7.692    1.736

X1            0.33     0.47     0.19
T-RATIO       2.16     3.04     1.01

X2                     0.60     1.55
T-RATIO                1.83     2.97

X3                              1.14
T-RATIO                         2.12

S            0.819    0.720    0.588
R-SQ         36.88    57.32    75.60
 MORE? (YES, NO, SUBCOMMAND, OR HELP)
SUBC> yes
 NO VARIABLES ENTERED OR REMOVED

 MORE? (YES, NO, SUBCOMMAND, OR HELP)
SUBC> no
--------------------------------------------------------------------------
```

The model chosen by the forward selection procedure uses X_1, X_2, and X_3 as the predictor variables.

(c) For backward elimination with *F-out* = **fremove** = 3.0, the Minitab command, subcommands, and output are as follows.

Minitab responds with:

```
--------------------------------------------------------------------------

MTB > stepwise c1 c2-c4;
SUBC> fenter=1000000.0;
SUBC> fremove=3.0;
SUBC> enter c2-c4.
```

```
 STEPWISE REGRESSION OF    Y     ON  3 PREDICTORS, WITH N =   10

    STEP          1         2
CONSTANT     1.7361    0.2626

X1             0.19
T-RATIO        1.01

X2             1.55      1.80
T-RATIO        2.97      3.85

X3             1.14      1.54
T-RATIO        2.12      4.15

S             0.588     0.589
R-SQ          75.60     71.41
 MORE? (YES, NO, SUBCOMMAND, OR HELP)
SUBC> yes
 NO VARIABLES ENTERED OR REMOVED

 MORE? (YES, NO, SUBCOMMAND, OR HELP)
SUBC> no

-----------------------------------------------------------------------
```

Thus, the final model chosen by the backward elimination procedure uses X_2 and X_3 as the predictor variables.

M7.5.1 First retrieve the file **pumpkin.mtp**. The data are in columns c1-c5. Create columns c11 c12 c13 using the `set` command. They should be as follows.

```
 C11   C12   C13
  1     4    16
  1     6    36
  1     8    64
  1    10   100
  1    12   144
```

Parts **(a)**–**(e)**, **(g)**, **(i)**, **(j)**, and **(k)** do not require Minitab calculations. For **(f)**, use the macro **growth** with column c20 given by

C20
0
0
1

and $1-\alpha = 0.90$. For (h), use the macro **growth** with column c20 given by

C20
1
8
64

and $1-\alpha = 0.90$.

In (**h**), the macro will give you a two-sided 90% confidence interval for $\mu_Y(8)$ in addition to the point estimate.

M7.5.2 First retrieve the file **ramus.mtp**. Parts (**a**)–(**d**), (**f**), (**g**), and (**h**), do not require Minitab computations. In (**e**), for a 95% confidence interval for β, use the macro **growth** with c11, c12, c13, and c20 given by

C11	C12	C13	C20
1	8.0	64.00	0
1	8.5	72.25	1
1	9.0	81.00	0
1	9.5	90.25	

and specify $1-\alpha = 0.95$. The result is $C[-4.292 \leq \beta \leq 11.065] = 0.95$. For a 95% confidence interval for γ, columns c11, c12, and c13 should be the same as before, but now use the values 0 0 1 (in this order) in column 20. The result is

$$C[-0.534 \leq \gamma \leq 0.351] = 0.95.$$

M8.2.1 The following commands will instruct Minitab to perform a weighted least squares analysis for the sulfur dioxide data (in the file **so2.mtp**) and display the results.

```
retrieve 'b:\so2';
portable.
let c3=1/(c2*c2)
name c3 'weights'
print c1 c2 c3
regress c1 1 c2;
weights c3;
xpxi m1.
print m1
```

All quantities needed to answer Problems 8.2.1–8.2.4 can be obtained from the Minitab output.

M8.2.2 The following commands will ask Minitab to perform a weighted least squares analysis for the soyburger data (in the file **soyburgr.mtp**) and display the results.

```
retrieve 'b:\soyburgr';
portable.
let c3=1/(c2**4)
name c3 'weights'
print c1 c2 c3
regr c1 1 c2;
weights c3;
xpxi m1.
print m1
```

Use the formulas in (8.2.24)–(8.2.29) to answer the questions. All quantities needed in these formulas can be obtained from the Minitab output.

M8.3.1 First retrieve the file **profsal.mtp** and then invoke the macro **theil** by issuing the command `execute 'b:\macro\theil'` . Relevant portions of the input and output from the macro are displayed below.

```
**********************************************************************

  Input a(0) and a(1), the coefficients of beta(0) and of beta(1),
  in this order.
DATA> 1 0

  Input the value of 1-alpha, the desired confidence coefficient
  for a two-sided confidence interval for theta.
DATA> .90

======================================================================

  The point estimate of theta is     28.00000


  For a two-sided confidence interval for theta with confidence
  coefficient equal to  0.875 (this is the value that is closest
  to the desired value of 0.900)
```

```
the lower confidence bound is      20.00000  and
the upper confidence bound is      31.11111
-----------------------------------------------------------------------
```

So we obtain $C[20.0 \leq \beta_0 \leq 31.1] = 0.875$. Note that the actual confidence coefficient of 0.875 is the closest possible to the desired confidence coefficient of 0.90.

M8.3.2 Use the macro **theil** again but use `a(0)` = 0 and `a(1)` = 1. Also, specify $1-\alpha = 0.90$. The results are

$$\hat{\beta}_1 = 2, \quad \text{and} \quad C[1.722 \leq \beta_1 \leq 2.375] = 0.875$$

M8.3.3 For (**e**) and (**g**), use the macro **theil** with `a(0)` = 1, `a(1)` = 0, and $1-\alpha = 0.90$. We get

$$\hat{\beta}_0 = 1.246 \quad \text{and} \quad C[0.364 \leq \beta_0 \leq 3.146] = 0.875$$

For (**f**) and (**h**), use `a(0)` = 0 and `a(1)` = 1. We get

$$\hat{\beta}_1 = 1.719 \quad \text{and} \quad C[1.075 \leq \beta_1 \leq 2.366] = 0.875$$

For (**j**), use the macro **theil** with `a(0)` = 1 and `a(1)` = 5. The result is $\hat{\mu}_Y(5) = 10.309$. For (**l**), the quantity of interest is $2.5\beta_1$. So use the macro **theil** with `a(0)` = 0 and `a(1)` = 2.5. Specify 0.90 as the desired confidence coefficient. The point estimate for $2.5\beta_1$ is 4.297 and a confidence interval is given by

$$C[2.688 \leq 2.5\beta_1 \leq 5.916] = 0.875$$

M9.3.1 Selected portions of the input and output from the macro **nonlin** are displayed below. The command to start this macro is `execute 'b:\macro\nonlin'` . Retrieve the data from the file **absorpt.mtp** before invoking the macro.

```
***********************************************************************

  Enter the number of the column containing the Y values.
DATA> 1

  Enter the number of the column containing the X values.
DATA> 2

  Enter the number of beta parameters.
DATA> 3
```

```
  Enter the number of the column for storing the initial values.
DATA> 3

  Enter the initial guesses for beta1, beta2, ..., etc., in order.
DATA> 0.8 -0.67 0.16

  Enter the number of iterations to be performed.
DATA> 20

  Enter the confidence coefficient for one-at-a-time two-sided
  confidence intervals.
DATA> .95

  +++++++++++++++++++++++++++++++++++++++++++++++++++++++++++++++++

  We have provided 4 nonlinear regression models in this macro.
  They are:

  Model 1: y = 1/(beta1 + beta2*x + beta3*x^2)

  Model 2: y = beta1*{1-exp[-exp(-(beta2+beta3*x))]}

  Model 3: y = 1/[1+exp(-(beta1+beta2*x))]

  Model 4: y = beta1+beta2*exp(-beta3*x)

  +++++++++++++++++++++++++++++++++++++++++++++++++++++++++++++++++

  Enter the model number ( 1 through 4 )
DATA> 1
====================================================================

  Parameter estimates and approximate standard errors (ASE)
  after the final iteration are:

        parameter          estimate                ASE

            beta1         0.8183216         0.06534496
            beta2        -0.6776059         0.06162592
            beta3         0.1632704         0.01450133
```

```
   For two-sided one-at-a-time confidence intervals for the beta
   parameters with confidence coefficient approximately equal to
    0.950 the lower and the upper bounds are given below.

         parameter        lower bound        upper bound

             beta1          0.6705009          0.9661422
             beta2         -0.8170135         -0.5381982
             beta3          0.1304661          0.1960746

   The estimated subpopulation standard deviation is    0.3696228


 Ready to continue ? (y = yes)
y

   The observed values (Y) and the fitted values (Yhat) are

     Observed          Fitted
        Y               Yhat

     1.800000         1.921834
     2.900000         3.289624
     6.000000         5.907684
     8.800000         8.606499
     6.600000         6.908622
     3.800000         3.922535
     2.900000         2.238325
     1.500000         1.388457
     1.100000        0.9299557
    0.5000000        0.6613533
     1.100000        0.4925096
    0.2000000        0.3801675

 Ready to continue ? (y = yes)
y

   A plot of the observed Y values against the X values and of the
   fitted Yhat values against the X values is given below.
```

```
   -
9.0+                 A
   -                 B
   -
   -                      B
   -                      A
6.0+            2
   -
   -
   -                           B
   -                           A
3.0+       2                        A
   -                                B
   -  2
   -                                     2    2         A
   -                                               2    B    B
0.0+                                                         A
     ------+---------+---------+---------+---------+---------+
          1.0       2.0       3.0       4.0       5.0       6.0
  A = Y vs. X              B = Yhat vs. X
```

In the above Minitab output, in order to save space, we have not shown the results that are printed out at the end of each iteration. You may get some warning messages from Minitab to the effect that some of the predictors are highly correlated with one another. For this problem however, this message can be ignored. Check that the results agree with those given in Exhibit 9.3.3.

M9.3.2 After retrieving the data from the file **coil.mtp**, invoke the macro **nonlin** with the command `execute 'b:\macro\nonlin'` . The inputs to this macro are similar to those in Problem **M9.3.1**, but choose model 2 for this problem.

M9.4.1 Choose model 3 in the macro **nonlin**. The initial guesses for β_1 and β_2 based on the linearization solution are -3.9627 and 174.755 respectively. Note that there are only two beta parameters in this model.

M9.4.2 For this problem you must choose model 3 in the macro **nonlin**. Note that there are only 2 beta parameters in this model.

M9.4.3 Use the macro **nonlin** and choose model 1. The initial guesses obtained from the linearization approach are $\beta_1 = 1.4026$, $\beta_2 = -1.2903$, and $\beta_3 = 0.2764$. The iterative process does not converge. The initial guesses obtained from the linearization approach appear to be unsatisfactory.

INDEX